D1451871

# An Introduction to Meteorological Instrumentation and Measurement

Thomas P. DeFelice
*University of Wisconsin at Milwaukee*

Prentice Hall
Upper Saddle River, New Jersey 07458

**Library of Congress Cataloging-in-Publication Data**

DeFelice, Thomas P.
An introduction to meteorological intrumentation and measurement
Thomas P. DeFelice.
p. cm.
Includes bibliographical references and index.
ISBN 0-13-243270-6
1. Meteorology—Measurement. 2. Meteorological instruments.
I. Title.
QC871.D34 1998 97-42092
551.5'028'4—dc21 CIP

Executive Editor: Robert A. McConnin
Total Concept Coordinator: Kimberly P. Karpovich
Art Director: Jayne Conte
Cover Designer: Bruce Kenselaar
Manufacturing Manager: Trudy Pisciotti
Production Supervision/Composition: WestWords, Inc.
Cover Photos: (Front) The TRMM satellite sensing the earth's atmosphere (Courtesy of Japan's National Space Development Agency (NASDA) and NASA). (Back) The Tropical Rainfall Measurement Mission (TRMM) Satellite with the conceptualized relative atmospheric coverage of each of its instruments (Courtesy of Japan's National Space Development Agency (NASDA) and NASA). PR indicates the precipitation radar sensor scan and ≈220 km swath width; VIRS denotes the visible and infrared radiometer sensor scan and >700 km swath width. TMI denotes the TRMM microwave imager sensor scan and ≥700 km swath width. The earth radiation sensor (CERES) and the lightning sensor (LIS) are located behind the two square meter PR antenna.



Reprinted with corrections January, 2000.

Printed in the United States of America

10 9 8 7 6 5 4 3 2

ISBN 0-13-243270-6

Prentice-Hall International (UK) Limited, *London*
Prentice-Hall of Australia Pty. Limited, *Sydney*
Prentice-Hall Canada Inc., *Toronto*
Prentice-Hall Hispanoamericana, S.A., *Mexico*
Prentice-Hall of India Private Limited, *New Delhi*
Prentice-Hall of Japan, Inc., *Tokyo*
Prentice-Hall Asia Pte. Ltd., *Singapore*
Editora Prentice-Hall do Brasil, Ltda., *Rio de Janeiro*

*I would like to dedicate this book*
*to my wife Michele; my sons Joseph and Matthew; my instrumentation instructors*
*Drs. A. Hogan, U. Czapski, G. Lala, G. Hill, A. Riordan, V. K. Saxena;*
*and to all past, present, and future meteorological instrumentationalists.*

# Contents

# Preface

The data acquired from measurements of meteorological parameters are used in a variety of activities that range from routine weather forecasting to specific problems encountered in cases such as airborne observation of cloud properties; relation of acid deposition in polar regions to climatic tendency; satellite detection of the ozone hole; or even the validation of satellite-derived meteorological parameters.

This book introduces the reader to the fundamentals of field measurements by providing the basics behind making meteorological or atmospheric environmental measurements using modern devices. The basics include: (a) how to choose an instrument; (b) how does an instrument make its measurement; and (c) how to ensure that the data is useable, and to what extent. The next step is an introduction to remote sensing platforms and to the meteorological measurements themselves. There is also a chapter on automated weather stations, such as ASOS, and a chapter on hydrometeorological measurements.

The text assumes that the problem to be solved and subsequent literature review of the scientific problem require one or more of the following measurements: radiation, temperature, pressure, windfield, moisture, cloud and precipitation amount and areal extent, evaporation, heat and momentum fluxes and their derivatives using mobile or stationary in-situ, or satellite platforms. Numerical exercises are provided at the end of every chapter. The appendices include fundamentals of electronics and analog-to-digital conversion. Also provided are a bibliography and glossary. The front and rear endpapers include a table of constants and a units conversion table.

The purpose of this book is not only to expose the reader to the different devices available, both present and future (where possible), but also to provide the reader with a general feel of how these devices physically obtain measurements, what to expect when making measurements with these devices, how to ensure the optimal performance of the devices, and more.

This book is intended to be an introduction to the measurement of basic meteorological parameters with modern and near futuristic devices for junior/senior-level students, and perhaps advanced sophomores, in Meteorological Instrumentation, Atmospheric Measurements I, Environmental Measurements I, Principles of Atmospheric Measurements I, or Environmental Instrumentation courses, etc. Such a course requires an introductory course in Atmospheric Science, Meteorology, or equivalent for respective majors. A third-semester physics course (usually electricity and magnetism–E&M) "with calculus treatment" might be beneficial and could be taken concurrently. Atmospheric Science, Meteorology, Environmental Science, Geography, Earth Science, and Geoscience departments (in order of likeliness) might offer an instrumentation course.

Presently there are no available texts that deal explicitly with meteorological measurements, especially those describing modern measurement techniques and devices. Most instrumentation texts that have been used are now out of print, or nearly impossible to obtain.

A recent survey of meteorological textbooks by the American Meteorological Society indicated that instructors use personal notes from their Meteorological Instrumentation courses. Actually, of the 58 schools that have a Meteorology or Atmospheric Science curriculum, only 26 offer an instrumentation class. The status of education and research on instrumentation and measurement systems in the early 1980s was poor (Brock, 1984), and it has not improved and may be even worse today. The lack of a standard instrumentation text has not helped the situation either.

The continued use of someone's notes, or books that are out of date, will only contribute to the continued and worsening problem of meteorological data being improperly made, misinterpreted, and misused. The use of personal notes has kept student cost to a minimum, but no two students take the same notes in a given subject from the same instructor, and most of them are very likely to permanently file them shortly after completing the course. It is these students who will be making or interpreting meteorological measurements now and in the future, and they will need a reliable, up-to-date reference.

This text assumes that individual instructors teach from their own course notes, or develop their notes from this text. It provides everyone, especially the student, with a readily available, updated, and comprehensible supplement that includes the basic theoretical framework of meteorological measurement and instrumentation. But remember, measurements are not theory, they are reality!

I would like to thank all those who have made a contribution to this text, in addition to its anonymous reviewers, who know who they are. Special thanks are extended to G. J. Carbone, University of South Carolina; D. Changnon, Northern Illinois University; U. Czapski, State University of New York at Albany; G. G. Lala, Atmospheric Sciences Research Center, State University of New York at Albany; J. M. Moran, University of Wisconsin at Green Bay; J. Song, Northern Illinois University; C. R. Stearns, University of Wisconsin at Madison; and A. Tokay, St. Louis University, for their constructive reviews of earlier versions of this book.

Special thanks are also extended to R. Hood and R. Spencer, National Aeronautics and Space Administration (NASA), Marshall Space Flight Center (MSFC), Global Hydrology and Climate Center (GHCC), Huntsville, Alabama; to M. Estes, Universities Space Research Association (USRA), Huntsville, Alabama; and to Dr. F. Six and NASA Joint Venture Initiative with University Educators program (JOVE, 144FG-46) for their contributions. The production-related efforts by the staff of WestWords, headed by P. McCutcheon, are also appreciated.

—Thomas P. DeFelice

CHAPTER ONE

# Fundamentals of Field Measurements

In this chapter, the reader is introduced to the underlying fundamentals of making meteorological measurements. We begin with a given problem that requires measurements, look at some of the questions that need to be considered when designing a field measurement campaign to solve the problem, learn about the performance characteristics of the measurement devices (at least at a fundamental level) and how to ensure that the performance of the instrument remains at its manufacturer-specified level. Finally, we explore how to retrieve the data, how to assure its quality once it is recorded, and how to estimate the errors that may cause your measurements to deviate from the actual value(s) being measured.

An experimental atmospheric scientist or meteorologist plans and then carries out a set of measurements that will provide the "true value" (i.e., simply the actual value) of the atmospheric conditions that are fundamental to answering a particular weather-related problem, or question. The weather-related problems or questions may be as basic as, what will the weather be like tomorrow, or they may be very specialized ones like the airborne observation of cloud properties, the relation of wet deposition in polar regions to climatic tendency, the radiative effect of cirrus clouds at the South Pole, or whether the globally averaged surface air temperature is warming or cooling. The problem defines the kinds of measurements that need to be made, as well as the sensitivity, accuracy, stability, and dynamic response required for each measurement.

The instrumentation likely to be used by the experimental meteorologist or atmospheric scientist include: radiation sensors primarily for determining the amount of incoming and outgoing solar and terrestrial radiation; electronic temperature sensors for air and possibly surface temperature measurements; barometers for air pressure measurements; wind vanes and anemometers or remote sensing devices for measuring the windspeed and wind direction; hygrometers or remote sensing devices for measuring the amount of moisture in the atmosphere (e.g., the relative humidity) and the surface; and precipitation gauges or remote sensing devices to determine the amount of precipitation that falls during a storm. Rawinsonde units are routinely used to obtain vertical temperature, moisture, and windfield (i.e., windspeed and wind direction) profiles. Lidars, radars, and radiometers are among the remote sensing devices that can be used to gather the same information as a rawinsonde as well as additional information on aerosols, clouds, and precipitation. Lidars, radars, and radiometers can be deployed on ground, airborne, and satellite-borne platforms.

Given a particular problem, the experimentalist asks questions like: What measured variables are required for this investigation? How frequent and over what period of time will the measurements need to be made? How accurate must they be? What instrument(s) will fulfill the measurement requirements? These very basic questions need to be answered before the experimentalist goes out into the field (or lab). The investigator must review the relevant literature and theoretical framework to aid in determining the variables that must be measured to solve the problem. The literature search for previous work done should go back at least 20 years, due to the periodicity in the funding of research areas. For example, acid rain research was widely funded in the 1980s, and there was very little funding before or after. The literature review will lead the investigator to the instrument(s) best suited for the task. The experimentalist is bound to advance the state-of-the-art pursuant to a particular problem. This includes everything from attacking new problems to approaching present problems from a different perspective to revisiting old problems with new, state-of-the-art measuring devices.

An instrument or measuring device generally consists of three fundamental components:

1. a sensing device (sensor);
2. a transducer, which translates a sensed quantity into either (a) a directly readable quantity (i.e., the position of the pointer on a scale or the meniscus of a fluid in a tube), (b) an electrical quantity (i.e., voltage, current, resistance, the frequency or amplitude of an electromagnetic wave), and occasionally (c) a count of audible clicks or the frequency, or pitch, of a sound wave; and
3. a transmission and/or readout device (i.e., an ohmmeter, voltmeter, amperemeter).

It senses a particular quantity with some degree of directness, ranging from none (or indirect) to complete (or direct). For example, obtaining the distance between the base and the top of the Empire State Building by measuring the change in barometric pressure using an altimeter is indirect, but using the pendulum period of a wire pendulum with the same length or using the time it takes a weight to fall from its top are more indirect. Using the vertical angle and distance from a theodolite is a little more indirect, but measuring the height of the Empire State Building with a large tape measure is direct. The air temperature near the surface of the earth is more directly measured with a mercury thermometer than it is with a satelliteborne remote sensor. The former requires the relation of the length of mercury column with air temperature, whereas the latter requires the computerized conversion of the total radiation in a particular wavelength band between the ground and satellite into the surface air temperature.

So there is more than one way to obtain the measurement of a particular quantity, especially if the quantity is air pressure or temperature. But which method do you use? The answer depends on your problem. But when answering the question of which instrument best satisfies the requirements, keep in mind that once the decision is made you will have to ensure that the instrument maintains an optimal performance throughout the investigation. This is accomplished by periodically calibrating the instrument before, during, and after its use in the field. The data is then analyzed and the problem solving begins. Let us discuss some of the more important characteristics that should be considered when choosing your instrumentation.

## 1.1. *Instrument Performance*

The choice of a meteorological instrument requires the consideration of the following (in order of perceived importance): the range, limitation, response, accuracy, resolution, representativeness, sensitivity, and precision of the instrument(s). There are a few additional factors that should be considered when choosing an instrument: compatibility with other instruments, simplicity of design, ease of reading, recording and/or transmission of data output, robustness, and durability, initial cost, and cost of maintenance. Cost, robustness and durability are of particular interest for most meteorological applications, and are actually a major concern to any investigation that requires continuous operation and complete, or partial, weather exposure, like most meteorological applications.

It is also important that the output signal of the chosen instrument cover the range of values that one might experience while making measurements. Where do the expected measurements fall with respect to the range of values measurable by the instrument? One wants to choose an instrument that will provide readings in the middle of its range, or wherever it is capable of giving its most accurate response. Otherwise the instrument is operating beyond its limitations. For example, most conventional cup anemometers fail to function well at windspeeds below 0.5 mph and over 100 mph, or in variable winds.

Differential equations are used to model the dynamic performance of measuring devices knowing that the models will never be exact. Linear differential equations with constant coefficients have well-known solutions that are easily inserted into computer models. Consequently, if the dynamic behavior of the measuring device may be described by linear differential equations with constant coefficients, then it is very easily studied before taking it into the field. Otherwise such equations are always approximations of the actual performance of measuring devices which often contain non-linear and time-variant performance characteristics.

The general dynamic performance model of the measuring device is discussed following Brock (1984) and is the linear ordinary differential equation

$$A_n d^nX/dt^n + A_{n-1} d^{n-1}X/dt^{n-1} + \ldots + A_1 dX/dt + X = X_I(t) \qquad (1.1)$$

where t is time and is the only independent variable, X is the dependent variable, $A_n$ are the equation coefficients or the measuring device or sensor parameters, and $X_I(t)$ is the forcing, driving, or excitation function.

This equation is ordinary because there is only one independent variable. Equation 1.1 is linear because its derivatives and the dependent variable only occur to the first degree. The latter excludes powers, products and trigonometric functions, such as sin X. The measuring device is considered time-invariant when its parameters or the coefficients in Equation 1.1 are constant.

The behavior of the measuring system that redistributes mechanical and potential energy as it responds to an input signal may be described by differential equations. For example, consider a moving mass, m, with a kinetic energy, KE, and a potential energy, PE. The PE would be zero if its position is at the ground state in a force field; for example, an object at sea level or altitude, $Z = 0$ m, has zero PE. Given that the PE is transformed into KE, and the differential equation is first order in velocity, then

$$m(dv/dt) + (D\,v) = F \qquad (1.2)$$

where v is velocity of mass, m, dv/dt is its acceleration, D is the dissipation factor, and F is an external force. Equation 1.2 applies for a weight initially at rest at top of the Empire State Building and suddenly dropped off, a cup anemometer, or the contraction (i.e., downward movement) of the mercury column after the maximum temperature is reached or in response to the onset of cold air advection at the site. If the measuring device has both PE and KE, then the differential equation describing its performance (i.e., Equation 1.2) becomes

$$m(d^2X/dt^2) + D(dX/dt) + (k\,X) = F \qquad (1.3)$$

where $(d^2X/dt^2)$ is acceleration, $(dX/dt)$ is velocity, v, and k is a potential energy parameter. Equation 1.3 applies to a wind vane responding to a change in windflow direction, and is a second order differential equation. Brock (1984) states that the order of the differential equation is always equal to the number of energy reservoirs. In a mechanical measuring device these energy reservoirs consist of stored kinetic energy elements and stored potential energy elements. Capacitors and inductors are energy storage elements in electrical systems, and thermal masses in thermal measuring devices, such as a liquid-in-glass thermometer.

The response of a linear measuring device is simply a sum of the responses from each of the various inputs sensed by the device, and is known as the *superposition principle*. The superposition principle allows one to determine the response to complex signals in the frequency domain by superimposing the responses to individual frequencies.

When the measuring device does not respond to an input, or there is no change in the distribution of the energy within the system it is said to be in a static state. The transition from one static state to another makes the device dynamic, and then its performance is described by a differential equation with respect to time, as mentioned above. What if a static characteristic of an instrument, such as the threshold of a cup anemometer, needs to be determined so that we can choose the best instrument to satisfy the measurement requirements? Then measurements of the output must be made for a number of different input values.

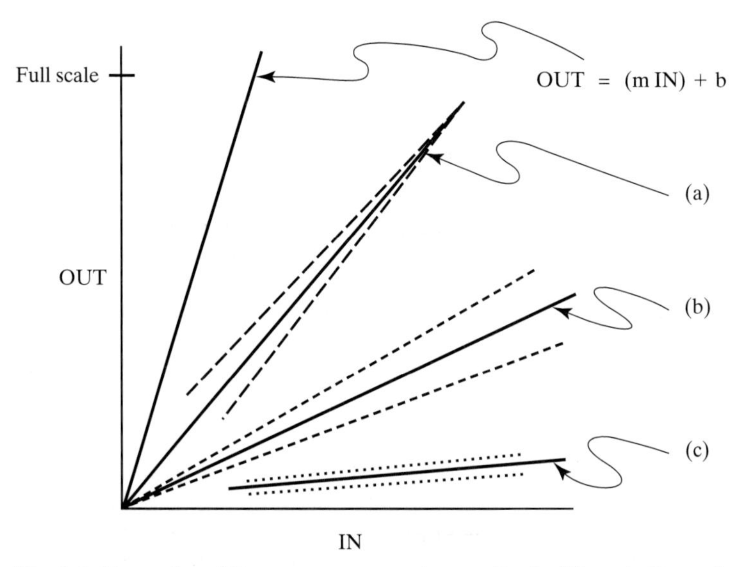


**Fig. 1.1** *Examples of linear responses to inputs. Dashed lines indicate the relative output accuracies for the respective curves.*

The response of the instrument is important for determining its accuracy. The smallest unit of accuracy is called the resolution of an instrument. It is normally desirable for an instrument to have an output that is linearly related to the input (i.e., output = m × input + b where m is the slope and b is the intercept), or a linear response to an input (Fig. 1.1). The linear response may be associated with a variety of output accuracies (Fig. 1.1), namely, an accuracy that is given by (a) OUT = ±c% of FULL SCALE (where c is a predetermined constant), (b) OUT = ±c% of the reading (input), and (c) OUT = ±c Units. For example, in (a), an instrument with this kind of response to an input yields very accurate readings at the top end of the scale. If our instrument has a linear response to an input (Fig. 1.1), and b = 0, meaning that the intercept of the linear equation is through the origin of our plot, c = 5 and full scale = 10 units, then the accuracy at full scale is 0.05 × 10 or 0.5 units. The accuracy of the output should be within 1 unit of the resolution. Digital instruments ideally should have errors within ±0.5 least significant bytes (L.S.B.), but practically it should be within ±1 L.S.B. For example a good digital voltmeter would have an output accuracy that is 0.01% reading ±0.005% Range at Full Scale ±1 digit. Thus, if the input is indicated as 6 volts (out of a 10 volt range full scale) the accuracy specification = [(0.0001 × 6) + (0.00005 × 10) + 0.0001] V = 0.0012 V = 1.2 mV.

Nonlinear responses (Fig. 1.2) do exist in some instruments, such as the dewpoint hygrometer and a hot-wire anemometer. Instrument systems with chart recorders, for example, require additional considerations with respect to system response, such as hysteresis, reading, and parallax errors. The response time of our chosen instruments must be able to resolve the natural variation of the respective measurements. The ability of an instrument to resolve the natural variation and the accuracy of that instrument usually coincide, but such is not always the case.

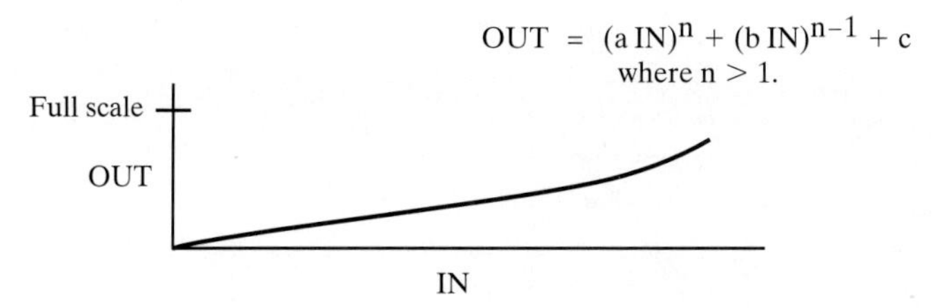


**Fig. 1.2** *Example of a non-linear response to an input.*

Other characteristics to consider include: the stability of the instrument, that is, does it yield repeatable measured values for a given input; the instrument's precision—low precision or high imprecision is related to a high standard deviation—and conversely; and the representativeness of the instrument's measured value. Representativeness is not an easily defined quantity. It has little to do with how well the instrument performs, but more to do with the spatial and temporal behavior of the parameter being measured, including the effects of the sampling platform. In other words, one might simply ask whether the "true" value obtained by the instrument is only valid for the immediate location or for a larger region. Single point measurements of atmospheric pressure and solar radiation are assumed to have spatial or temporal representativeness over very large spatial areas, or very long times. In contrast, single point measurements of turbulence, winds, temperature, and humidity are not. What about precipitation? Consider the type of precipitation, intensity and duration, the effect of topography, and gauge exposure (i.e., its height, surrounding obstructions), before answering. Representativeness can be quantified through the use of autocorrelations.

Naturally, the output signal of our instrument has to be sensitive enough to be distinguishable from its corresponding noise level, if the signal-to-noise ratio is specified. Note that a highly sensitive instrument system is very expensive and one needs to choose the sensitivity according to their operational requirement. High resolution and high accuracy are often desirable and expensive, but difficult to achieve simultaneously in one instrument. Trade-offs are necessary.

***1.1.1. Response Time.*** The response time of an instrument has also been termed time response, time constant, time transit, speed of response, time lag coefficient, lag coefficient, or the time lag. These terms actually have subtly different definitions despite their interchangeable use. The term *response time* will be adopted in this text, and is taken to indicate the time it takes an instrument to completely respond to a changing input. In contrast, the time constant simply corresponds to the time it takes for the instrument to reach 63.2% of the response to a step input change, and is discussed below.

The determination of the instrument response time is done in a controlled environment (laboratory). The procedure used to determine the magnitude of the response time of an instrument requires mathematical or graphical manipulation. The mathematical determination of the response time for a first order measurement system will be illustrated using the response time for liquid-in-glass thermometers.

Second order systems, which may have a damped oscillatory response, together with sinusoidal and random forcing functions, are left as exercises in the appropriate chapters. Brock (1984) has additional details that might prove helpful.

The liquid-in-glass thermometer takes time to respond to a change in the temperature it senses. As a result, the indicated value usually represents a period other than the instant when the reading was made, and the lag is related to the time constant of the thermometer. Actually the lag depends on the thermal properties (i.e., material, dimensions) of the temperature sensor, and the conditions of the surrounding medium (i.e., windspeed, natural temperature variability, whether the sensor is shielded and ventilated). Knowledge of the magnitude of this lag is important in understanding the limitations of the particular temperature measuring device, and is normally found through laboratory experimentation.

However, how a temperature device responds to a thermal change within the medium it senses can be illustrated by assuming that the device has a first order response. A first order response suggests that the rate of change in output (i.e., response) is proportional to the difference between the output and input. Thus there must be a change in heat flow from the thermometric device in order to change the temperature it senses. Suppose a mass, M, of air flows past a thermometric sensor in unit time and as a result the air's temperature changes from $T_\infty$ to T. Actually a larger mass of air than M will be affected by the thermometric device, but the temperature changes will not equal $T_\infty - T$ for all of that mass. The heat loss, or gain, by the air in a short time, dt, is $Mc_p(T_\infty - T)\,dt$, where $c_p$ is the specific heat of the air at constant pressure, and is the amount of heat available for absorption by the thermometric device with thermal capacity C. The absorption of the available heat by the thermometer will change its temperature by an amount equal to dt. Thus, using previously defined symbols and assuming that convection is the only mode of heat transfer to the device, and that the device has no internal resistance to the heat flow, then the heat balance equation takes the form,

$$-MC_p(T - T_\infty)\,dt = C\,dT. \tag{1.4a}$$

Solving for dT/dt yields

$$dT/dt = -\tau^{-1}(T - T_\infty) \tag{1.4b}$$

where $\tau$ is the time constant of the temperature measuring device and is given by

$$\tau = C(Mc_p)^{-1}. \tag{1.4c}$$

Assuming that the sensor receives a Step input (such as a common test signal used in many laboratory procedures) then:

step

$T_0$ ⎯⎯⎯⎯⎯⎯⌐⎯⎯⎯⎯⎯⎯ $T_\infty$

$$T = A\,EXP[-t/\tau] + B$$

when $t = 0, \quad T = T_0 \quad A = T_0 - T_\infty$

when $t = \infty, \quad T = T_\infty \quad B = T_\infty.$

**Table 1.1** *Percent response completed versus elapsed time for a temperature sensor receiving a step input*

| *% Response completed* | *Elapsed time* |
|---|---|
| 50.0% | $0.7\tau$ |
| 63.2% | $1.0\tau$ |
| 90.0% | $2.3\tau$ |
| 95.0% | $3.0\tau$ |
| 99.0% | $4.6\tau$ |
| 99.5% | $5.3\tau$ |
| 99.9% | $6.9\tau$ |

Rearranging, integrating, and simplifying yields

$$T = (T_0 - T_\infty)\mathrm{EXP}[-t\tau^{-1}] + T_\infty. \tag{1.5a}$$

Further simplification gives

$$\mathrm{EXP}[-t\tau^{-1}] = (T - T_\infty)(T_0 - T_\infty)^{-1}. \tag{1.5b}$$

Table 1.1 is constructed to help visualize how long it would take the temperature sensor to reach the equilibrium (ideally "true") air temperature. When $t = \tau$, the temperature difference between the sensor and its surroundings has decreased to $e^{-1}$ (i.e., 1/2.718 ...) of its initial value, or

$$(T - T_\infty)(T_0 - T_\infty)^{-1} = \mathrm{EXP}[-t\tau^{-1}] = \mathrm{EXP}[-1] = 0.368. \tag{1.5c}$$

Thus when the time is $1.0\tau$ after the sensor was placed in the environment, or 1 time constant, the sensor's temperature will indicate 63.2% of the equilibrium air temperature. Table 1.1 suggests that it takes at least 5 time constants for the sensor to be within 99% of the equilibrium temperature. Covering the sensor with an insulating material will also introduce a lag into its response. The time constant may also be determined by graphical means.

Given that the air temperature is changing at a constant rate of $\beta$ degrees per second (sometimes termed a "RAMP" input) during measurement, then the indicated value of the sensor is less than the actual air temperature value by an amount $\beta\tau$ (Fig. 1.3). Note that the "RAMP" input is a more realistic description of the type of signals encountered in many meteorological processes.

Consider the scenario when the temperature of the medium undergoes a simple harmonic variation with an amplitude of $T_{mn}$ and period P seconds, sometimes termed a sinusoidal input. The resulting percent response versus time is highlighted in Table 1.2. This scenario and the previous one if applied to liquid in glass thermometers neglect temperature differences inside the bulb itself.

Recording devices also have time constants. The larger time constant in a measurement system (e.g., sensor + recording device) will dominate. Quick time responses are usually desired, but a relatively long time constant may be acceptable

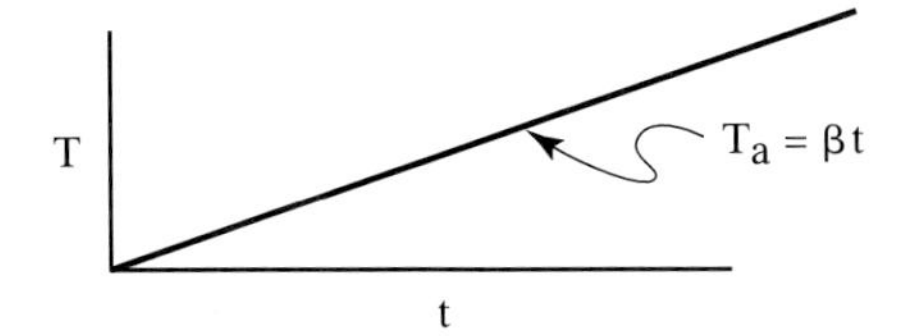


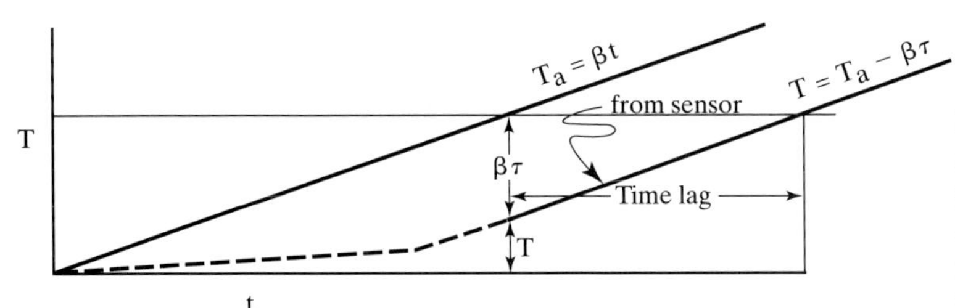


**Fig. 1.3** *Electronic temperature sensor response to air temperature changing at a constant rate. $\tau$ approaches time lag when elapsed time, t, is much greater than $\tau$.*

**Table 1.2** *Percent response versus time for a temperature sensor receiving a sinusoidal input*

| $T/T_{mn}(\%)$ | $\tau/P$ | $P/\tau$ |
|---|---|---|
| 99.0 | 0.016 | 62.5 |
| 90.0 | 0.078 | 12.8 |
| 70.0 | 0.186 | 5.4 |
| 50.0 | 0.275 | 3.6 |
| 15.7 | 1.000 | 1.0 |
| 5.0 | 3.190 | 0.313 |

depending on the requirements (Fig. 1.4). For example, one might choose a longer time constant to get a data record that is easy to work with, i.e., for diagnostic purposes. There might even be a need to use or design a specialized device to measure the time variation rather than the mean or instantaneous value of some quantity. A simple example would be rainfall rate versus total rainfall over a time interval.

Keep in mind that the standard meteorological variables—i.e., temperature, pressure, humidity, windspeed and direction, concentration of particles and gases—are instantaneous values derived from a balance of forces or fluxes in our ever-changing environment in which the instrument itself may be a determining factor. Take temperature as an example. The length of a mercury column in a given tube depends primarily on the expansion of the mercury as well as of the glass if the sensor is at radiative equilibrium with its environment (i.e., all incoming heat fluxes balance all outgoing heat fluxes). If the air temperature of the environment surrounding the thermometer is the desired quantity, then the convective–conductive heat flux from the air to the thermometer must determine the temperature reading. In order for that to happen, the mass and heat capacity of the thermometer has to be small

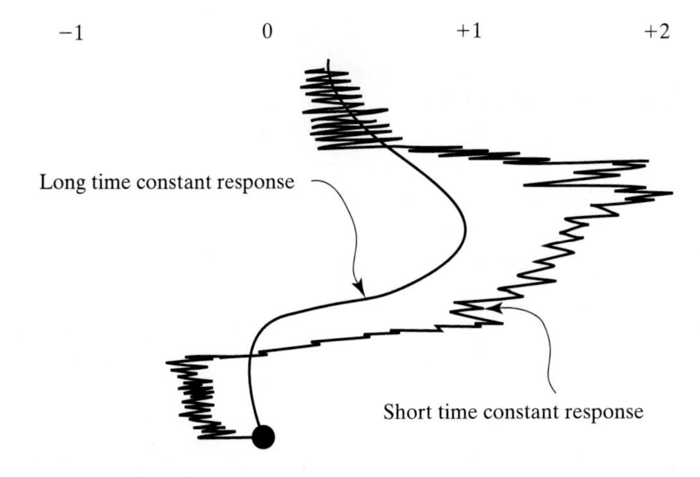


**Fig. 1.4** *An example of a device with a short time constant and one with a longer time constant.*

compared to the mass and heat capacity of the air surrounding it. The mass of air is approximately 1 g per liter, while that of a conventional thermometer is approximately 100 g. Therefore in order to be "small" we need at least 1000 liters of air to come into contact with the thermometer within the response time (i.e., ≈12 min). This is the reason why aspirated thermometers are often used in meteorological practice. But we still have to rely on the uniformity of temperature of the aspirated air. If we go to higher levels of the atmosphere, where the density may decrease to 0.01 or 0.001 of its surface value, then the problem becomes unsolvable if conventional thermometers are used—particularly because the radiative fluxes (solar flux absorbed by the thermometer surface and outgoing blackbody emission) that were easily neglected near the surface using reflective shields become very significant.

## 1.2. Instrument Calibration

Assume that our instrument is chosen and either purchased, borrowed, made by, or adapted from your own or someone else's lab. We must always make sure that our instrument operates according to manufacturer specifications before and during field use. This involves a series of steps known collectively as an instrument calibration, and involves: (1) providing the instrument with a known standard input of a previously established accuracy and recording its response (i.e., a **primary standard**—see glossary); and (2) comparing the response to the equivalent standard response derived from the factory for that instrument.

The actual standard input might be a known current, voltage, light intensity, relative humidity, windspeed, temperature, pressure, number of given sized beads of a specific particle density, chemical content, index of refraction, etc. The measurement of length has a nonvariant standard that can now measure lengths to $1 \times 10^{-9}$ m, according to the National Institute of Standards and Technology (NIST). A piston gauge is the primary standard for pressure and it can determine pressures from 0.01 to 10,000 psi with a resolution and calibration uncertainty in the parts per million range. Mercury barometers are commonly used as **secondary or reference standards** (see glossary). A platinum resistance thermometer is used as the primary standard for temperature. A laser doppler anemometer is the primary standard for low windspeed measurements. The primary standard for humidity is a gravimetric method of the absorption of water vapor mixed with a gas by a dessicant and the increased mass is measured. The primary standard for measuring solar, terrestrial, and atmospheric radiation is an absolute cavity pyrheliometer. Additional details on meteorological measurement standards are provided by Brock (1984).

When you are out in the field away from a lab, you must still perform the same calibrations. The calibration should be checked periodically in order to ensure that each instrument maintains its optimal performance. The periodic check that an instrument is operating at its optimum level is often termed *quality assurance*, and is discussed later.

The instrument that is calibrated should posses long term stability with respect to its intended period of use. That means, if the instrument is calibrated at the beginning of your field period, then its calibration at the end of the field period should be the same. A record of calibration over the lifetime of an instrument is known as its calibration history. A calibration history will show the deviation of an instrument's response to a known standard input compared to its very first response to the same standard input. Note that the aforementioned deviation is known as *instrument drift*. Calibration removes the systematic error associated with the particular sensor (Brock, 1984).

## 1.3. Data Acquisition

Now that we have our choice instrument, and it is operating correctly, we are ready to take some measurements. The process of making measurements is simply termed data acquisition, and it requires some knowledge of electronics (see Appendix A). Those with little background in electronics are encouraged to examine other sources on this topic. The exercises at the end of this chapter will help you practice some of the electronic basics. Please note that two or more of those sources might use different terminology for the same electronic device.

Each instrument does not necessarily sample information at the same rate. The rate at which each instrument samples and how quickly it takes each instrument to respond to a change in the parameter it is measuring (i.e., its response time) are very important questions that must be considered, especially if you are trying to store data on a computer or a datalogger (Fig. 1.5).

***1.3.1. Timing of Measurements.*** When storing these sensed data on a datalogger or computer you need to know how frequently the instrument sends data to the data acquisition system, as well as how much storage space you have. This will allow you to optimize your data collection. So how often should one sample a sensor to obtain all of the information it detects? (That is, how often should a sensor be queried to retrieve all of the information it senses.) Let's assume that the analog sensor signal, V, has a range $V_{max} - V_{min}$ and a resolution $\Delta V$. Then the number of possible states is

$$n = [(V_{max} - V_{min}) \Delta V^{-1}] + 1. \quad (1.6)$$

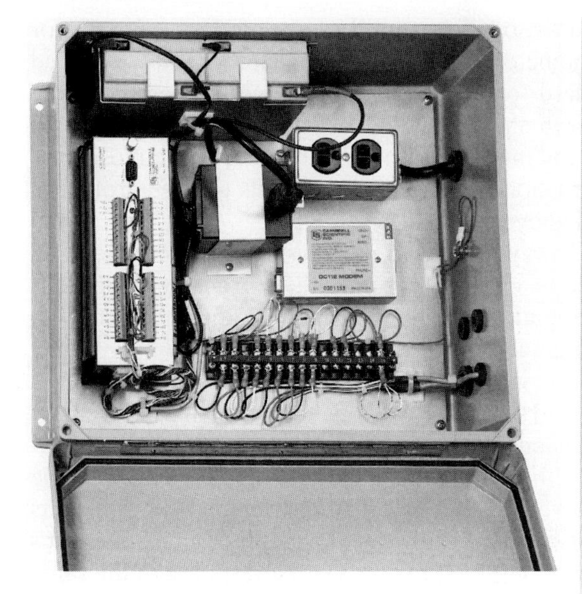


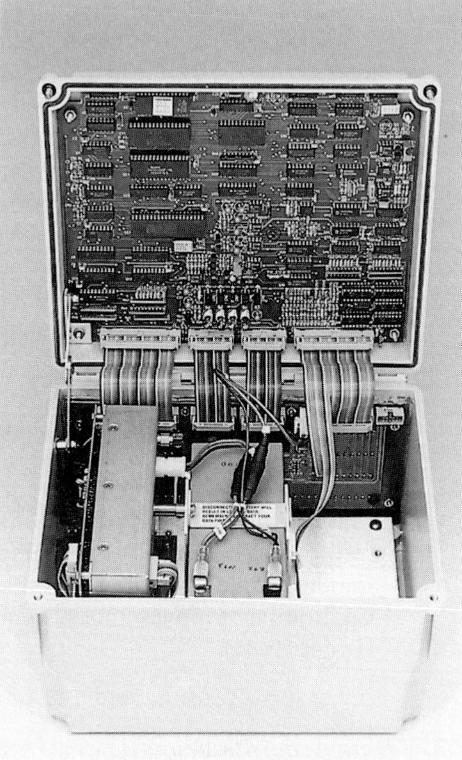

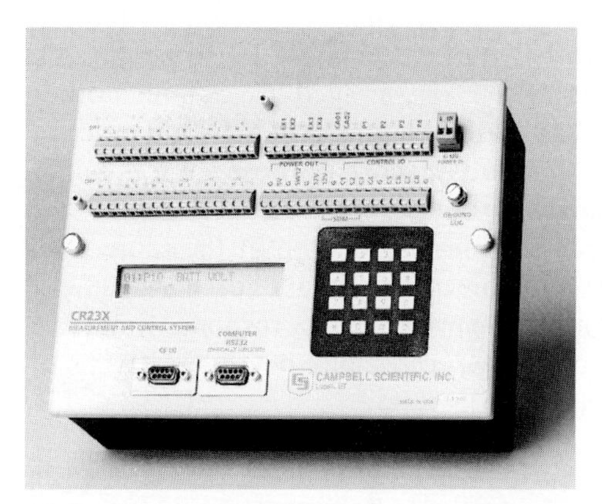


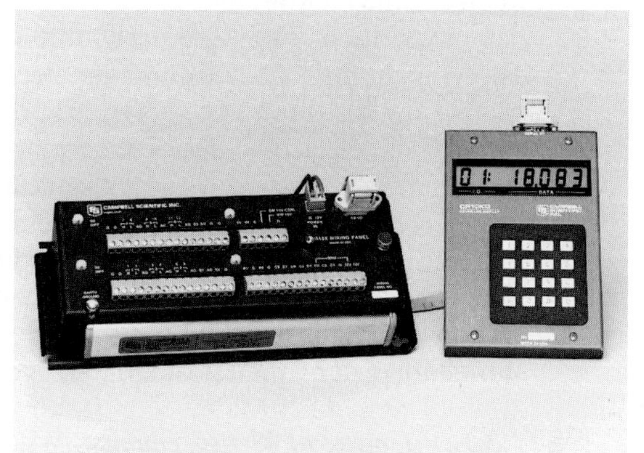


**Fig. 1.5** *Some examples of dataloggers. (Courtesy of D. Gilmore and D. Katz of Climatronics Corp., K. Schlichting of Handar, and J. Goalen of Campbell Scientific, Inc.)*

If a binary analog-to-digital converter, ADC, (see Appendix B) is used, then $\Delta V$ would correspond to one binary bit and n would be the number of possible states for the digitized analog signal. If the system is first order, then the signal changes at a rate dV/dt, where

$$dV/dt = -\tau^{-1}[V(t) - V_{environment}], \tag{1.7}$$

and the time for the signal to change an amount equal to $\Delta V$, or to change one binary bit, $\tau$, is given as

$$\tau = \Delta V (dV/dt)^{-1} \tag{1.8}$$

where,

$$\Delta V = V_{environment} - V(t) = \tau[(V(t + \Delta t) - V(t - \Delta t))/(2\Delta t)], \tag{1.9}$$

and $\Delta t$ is sample time interval. For example, if one samples 10 times during each time constant then $\Delta t$ is $\tau/10$.

The maximum value of dV/dt would correspond to the smallest time interval between readings. If the environmental value varies in a sinusoidal manner, then, according to Fuchs (1971), the sampling interval between successive samples should be $(1 + w^2\tau^2)^{-0.5}$ times $\pi$ times $\tau$, where $w = 2\pi/p$, and p is the period. Consequently, the time for a first order measurement device to just detect a change in signal, or the time period for sampling, $p_{sample}$, may be expressed in terms of the full scale range, resolution, and the time constant of the measuring device, $\tau$, as follows:

$$p_{sample} = \{4\pi^2\tau^2[(V(\Delta V)^{-1})^2 - 1]^{-1}\}^{0.5}. \tag{1.10}$$

Fritschen and Gay (1979) provide some additional background for Equation. 1.10.

---

**Example 1.1**

If a first order device has a $\Delta V/V = 1/1000$, and a time constant of $\tau$, then how long should the sample duration be? $p_{sample} \approx \tau/159$ and is actually an underestimate of $p_{sample}$.

---

Given that our sensor has a frequency response, $f_{device}$, then, according to Shannon's Sampling Theorem (e.g., see Fritschen and Gay, 1979), if you sample at the rate greater than or equal to $2f_{device}$, it will be possible to reconstruct the response of the instrument without error. Simply, this means that you need at least two samples during the time associated with the highest frequency response of the device. There are two conditions for reconstructing instrument response:

1. The signal behaviour remains constant during $p_{sample}$, and $p_{sample} \gg (f_{device})^{-1}$.
2. The input frequency response must be less than $f_{device}$.

If a higher frequency than $f_{device}$ is present, then sampling at $2f_{device}$ will result in an apparent signal at a lower frequency. The minimum sampling frequency to permit reconstruction of the device's signal equals $2f_{device}$, and has been termed the Nyquist frequency. Sampling at a frequency less than the Nyquist frequency results in aliasing. Aliasing is seldom detected in standard meteorological measurements unless the measurements are used for spectral analyses and other turbulent transfer phenomenon.

## 1.4. Quality Assurance of Data

We must first learn how to operate and calibrate any instrument correctly before we use it in the field. This is best accomplished by attending centralized training centers that specialize in the instrumentation that you will use. Once we have learned how to operate and calibrate our instruments, deployed them in the field, and know that they are otherwise ready for data collection, we may begin collecting data. Our data collection campaign should (i) follow a standardized operations plan, (ii) include routine checks of the instrument's calibration using field or ideally secondary standards (see glossary), and (iii) deploy multiple numbers (at least 2) of each measuring device at the data collection site (i.e., co-locate your instruments) in order to ensure optimum performance throughout the measurement campaign. Optimum performance of our instrumentation yields measured values that are known to be very near "true". A standardized plan is essential if more than one site is involved, or if our measurements are to be compared with similar studies.

The items just mentioned in this section collectively form the skeleton of what is termed a quality assurance (QA) plan. The QA plan should describe the purpose and design of the project, and outline the objectives, measurement and analysis methods, and milestones of the project. The latter must leave a clear understanding of the overall scope of the project goals and activities. All individuals involved with the project should be aware of their responsibilities, and one of them should be charged with the management of the QA program, including the preparation of status reports, and evaluating the effectiveness of the QA plan. The QA plan should specify the required accuracy, precision, and completeness (ratio of actual to total possible observations) for each measured parameter.

Naturally it is helpful to know the objective of the project and hence the level of accuracy and precision required for each parameter before constructing the plan. Completeness specifically applies to continuous measurements over an extended period of time. It should also include the objectives for representativeness, and field intercomparisons with different instrumentation that measure the same quantity (or comparability). Comparability allows for standardization of future measurements by these devices, and perhaps improves the way some of them make measurements. The plan should also contain a description of all instrumentation (i.e., in the measurement system and methods section) that includes: a theory of operation involved in measuring each parameter, methods or guidelines for selecting a site and locating sensors, precautions to guard against unwanted sensor inputs, sensor signal processing and recording methods, response times and sampling rates.

Detail the procedure for calibration of each measuring device, and keep a record of when the calibration was performed, and the results of each calibration. The standards used must be kept in calibration against primary standards, and should be traceable to NBS or other fundamental standards. The plan should also describe the internal and external performance and systems audit procedures. **Audits** (see glossary) should occur when the measurement program begins and at regular intervals thereafter.

The QA plan must contain a detailed description of data validation procedures and criteria including methods of handling missing data or will be considered questionable or invalid. Data validation is a non-glamorous, but extremely important editing process in which measured data are screened in comparison to an expected value or range of values. A plan to implement corrective measures in the event data quality deteriorates beyond acceptable limits must also be included. U.S. EPA (1976) discuss procedures for corrective action.

The QA plan should also specify the schedule and content of the performance reports of the measurement system and provide quantitative information on the quality of the data produced by the program. These reports should include: an assessment of data accuracy, precision, and completeness; the results of performance and system audits; and a description of any significant QA problem, along with recommended solutions and corrective actions taken.

There are many quality assurance programs available and they are all instrument, site and task specific. It is beyond the scope of this text to highlight the details of these quality assurance programs. Instead, suffice it to say that one should ideally check the performance of all of their measuring devices on a daily basis. This way the optimum performance of each instrument is ensured and your data return per cost ratio is high, indicating a successful field campaign. If it is not possible or practical to check the performance of your instruments on a daily basis (i.e., because of ongoing operations), then it should be done on the first available opportunity. The performance checks *should ideally not be more than 2–4 weeks apart.*

The frequency of calibration depends, of course, on the design of the instrument. An ultrasonic anemometer needs to be calibrated before and after each use, whereas a simple rainguage does not need frequent recalibration. Finkelstein et al. (1983) provides details of performance and systems audits, and data validation procedures for meteorological measurement systems.

Even though modern day computers with the appropriate software are replacing chart recorders, it is still wise to have or to perform periodic reliable data backups (i.e., use a chart recorder, a few boxes of diskettes, and/or a Polaroid camera, making sure you have enough film for your data backup) during any measurement campaign, especially one outside of the laboratory. *The important point is to back up the data.* A word to the wise, though, taking one or more backups into the field could cause some extra work. However, it is better still to relax later, and have useable data.

## 1.5. *Measurement Errors and Their Estimation*

Once measurements are made and converted to a legible form of "true" values for analysis, the next step is to determine how well the measurements represent the "true value" of the parameter being measured (see glossary). Deviations from the "true value" generally fall into at least one of two categories, namely, (i) systematic error, or (ii) random error. Errors caused by the operator may either be systematic or random and come in many different variations, i.e., parallax, other observation errors, failure of calibration, exposure.

Operator errors are probably the most costly of the three, since the only way to recover the data is to repeat the measurements in a more careful manner. Operator errors are greatly reduced in frequency and magnitude by practicing the sampling procedure a few times before arriving in the field. It is also advisable to run through the sampling procedure a couple of times in the field before making the measurements. Remember your prime and only directive when in the field is to make measurements that are as close as possible to the true value of the quantities you are measuring.

Systematic errors occur because of faulty system-wide methods such as a defective measuring apparatus, incomplete working equations, or, in some cases, the environment (Wall et al., 1972). Some examples of systematic errors include: (a) making length measurements with a plastic ruler that has a melted scale, (b) making mass measurements with a non-zeroed "scale" and (c) making radiation or optical measurements when the sun is at a low angle causing a consistent error. These errors are corrected by: (a) replacing or correcting the method of measurement, (b) calibrating the measuring apparatus, or (c) adjusting the working equations. Systematic errors cause every measurement to be off by a similar magnitude, and they are usually more significant than random errors.

Random errors are present in nearly all measurements, and they arise from sudden or uncontrollable changes within the measured environment that are significant enough to affect the measuring device, and the quantity being measured (Wall et al., 1972). They may also be caused by a random careless act of an observer or his/her associate. Any quantity that exhibits the aforementioned properties is termed a random variable, and it may be described statistically by a normal (or Gaussian) distribution (i.e., the well-known bell-shaped curve). This implies that these errors are equally likely to be positive or negative and relatively small. The effect of random errors are minimized by taking a large number of measurements of the particular quantity and using the expected value of these measurements as the best estimate of their "true value."

A frequent and important cause of both random and systematic errors may not be in the sensing device, but in the electronic circuitry of the transducer, transmitter, or readout. Induction by random or systematic electromagnetic fields is a common cause.

The estimation of the aforementioned errors inherent in the measured quantities is necessary and may be accomplished as outlined in the following pages.

*Systematic Error Estimation.* There are two general ways to estimate the systematic error of a dependent quantity. They simply involve looking at the error of each measurable variable that the dependent variable is related to. The two methods are the Determinant-Error Method (e.g., Wall et al., 1972) and the Log Derivative Method.

*Determinant-Error Method.* Suppose that the dependent quantity T (to be determined) is related to two (for simplicity) independent measurable quantities Q and Y by the equation,

$$T = f(Q,Y) \tag{1.11}$$

where f is a known function of the independent variables Q and Y, and T is the dependent variable. Equation 1.11 is valid for measured and "true" values. Let $Q_0$, $Y_0$, $T_0$ designate "true" values, and Q, Y, and T represent measured values. Then Equation 1.10 may be written in terms of the "true" values as

$$T_0 = f(Q_0, Y_0). \tag{1.12}$$

An expression for the errors is obtained by subtracting Equation 1.12 from Equation 1.11 yielding:

$$(T - T_0) = f(Q, Y) - f(Q_0, Y_0). \tag{1.13}$$

Given sufficiently small values of $(Q - Q_0)$ and $(Y - Y_0)$, the right side of Equation 1.13 may be approximated by a linear function of these small quantities, given by Taylor's Theorem. Assuming this approximation is valid, then

$$(T - T_0) = \frac{\partial f}{\partial Q}(Q - Q_0) + \frac{\partial f}{\partial Y}(Y - Y_0). \tag{1.14}$$

The partial derivatives $\partial f/\partial Q$ and $\partial f/\partial Y$ represent the respective rates of change of T with respect to Q and Y. They are typically evaluated at $T_0$, $Q_0$, and $Y_0$. However, for small intervals they may be evaluated anywhere in the interval, provided that the function varies slowly. $(T - T_0)$, $(Q - Q_0)$, and $(Y - Y_0)$ are the respective errors by definition (note that the error must be added to the "true" value to obtain the measured value). The correction is defined as the negative of the error. Equation 1.14 shows how to compute the errors in T given the errors in Q and Y, and it is normally called the *determinant-error equation* for Equation 1.11. It is customary to simplify Equation 1.14 as follows:

$$\Delta T = \frac{\partial f}{\partial Q}(\Delta Q) + \frac{\partial f}{\partial Y}(\Delta Y) \tag{1.15}$$

where $\Delta T = T - T_0$, $\Delta Q = Q - Q_0$, and $\Delta Y = Y - Y_0$. Note that in this form it is obvious that the error equation, Equation 1.14, is just the differential of Equation 1.11. Since a systematic error is definite in sign and magnitude, a known systematic error in Q can be completely represented by $\Delta Q$, and that in Y by $\Delta Y$. These could then be used in Equation 1.11 to compute the systematic error in T. The "true" value of T would then be $T_0 = T - \Delta T$ (recall T is the indirectly measured value). An alternative procedure is to compute $Q_0 = Q - \Delta Q$, and $Y_0 = Y - \Delta Y$ first, and then use Equation 1.12 to compute $T_0$. Either way should yield the same value of $T_0$, but only if Q and Y are measured independently of T.

*Log Derivative Method.* Here we begin with a function such as that in Equation 1.11, namely, $T = f(Q, Y)$. Then take the natural log of both of its sides, and then put the resulting expression into its differential form. For example, consider the following:

$$\begin{aligned} R &= VI^{-1} \text{ (or } V = IR) \\ \ln R &= \ln V - \ln I \end{aligned} \tag{1.16}$$

$$\frac{\partial R}{R} = \frac{\partial V}{V} - \frac{\partial I}{I} \quad \text{or in terms of finite sums} \quad \frac{\Delta R}{R} = \frac{\Delta V}{V} - \frac{\Delta I}{I}$$

The maximum error estimate of the resistance is obtained by adding the magnitudes of the errors of the independent variables regardless of the log-derived expression. Consequently,

$$\text{Max.}\frac{\Delta R}{R} = \left|\frac{\Delta V}{V}\right| + \left|\frac{\Delta I}{I}\right| \tag{1.17}$$

where, $|\ |$ represent the absolute values of the variable within them.

---

**Example 1.2**

A multimeter that is capable of measuring voltage, V, and current, I, with an accuracy of ±0.01 volts, and ±0.01 milliamps, is used to measure the voltage and current in a simple circuit. The results of these measurements yield

$$V = 0.50 \text{ volts; } I = 0.70 \text{ milliamps} = 0.70 \times 10^{-3} \text{ Amps.}$$

What is the resistance, R, of this circuit? What is the maximum error in R?

$$R = VI^{-1} = 0.50\,v\,(0.70 \times 10^{-3}\,\text{Amps})^{-1} = 714 \text{ Ohms [actually 714.29]}$$

$$\text{Max. error in R} = \frac{0.01 \times 10^{-3}\text{ Amps}}{0.70 \times 10^{-3}\text{ Amps}} + \frac{0.01\text{ volts}}{0.50\text{ volts}} \approx (0.034)\,714.29\text{ Ohms}$$

$\Delta R = \pm 24$ Ohms [24.29]. Note: It is highly improbable that the error in R is this large.

---

*Random (Probable) Error Estimation.* The following assumes that the measured values have been corrected for all systematic errors, and only the random errors remain. Random errors in the directly measured quantities Q and Y will produce random errors in the computed quantity T. This implies that the computed values of T will have a distribution with its own "true value" and standard deviation. Here we also assume that T is related to Q and Y as given in Equation 1.11. The estimated random error in T based on the random errors in the directly measured quantities that produce T is commonly expressed as

$$(\Delta T)^2 = \left(\frac{\partial f}{\partial Q}\right)^2 (\Delta Q)^2 + \left(\frac{\partial f}{\partial Y}\right)^2 (\Delta Y)^2 + 2\frac{\partial f\ \partial f}{\partial Q \partial Y}(\Delta Y)(\Delta Q) \tag{1.18}$$

where $\Delta Y$, $\Delta Q = \pm$ the standard deviations of the respective variable divided by the square root of the number of measurements minus 1. Recall that the variance is the standard deviation squared; and the standard error (dispersion of the means) is the standard deviation divided by the square root of the number of measurements.

---

**Example 1.3**

Consider the same example that was used earlier for the systematic error case, i.e., $R = V I^{-1}$. What is the random error in R?

$$\Delta R = \{[\partial R/\partial I)\Delta I]^2 + \left(2\frac{\partial R}{\partial I}\frac{\partial R}{\partial V}\Delta I \Delta V\right) + [(\partial R/\partial V)\Delta V]^2\}^{0.5}$$

$$\Delta R = \{[-V\Delta I I^{-2}]^2 + (-2VI^{-2}I\ ^{1}\Delta I\Delta V) + [\Delta V I^{-1}]^2\}^{0.5}$$
$$\approx 5 \text{ Ohms (using the same values as above)}$$

The result is not as bad as the maximum error, as expected. Generally, $\approx$ 80% of readings should fall well within the maximum error, and 68% of readings should fall within the random (or probable) error.

---

Now that we have an idea of how to go about determining the errors of a given quantity, what would be the errors in some quantity, say temperature, if measured by the system shown in Fig. 1.6. The task is left as an exercise.

*Mixed Error (Error Intervals or Uncertainties) Estimation.* The significant errors in a measured system are seldom purely systematic or purely random. They are usually a mixture of both types of errors. The presence of unspecifiable systematic and random errors that influence a measurement is termed *mixed errors*. One may determine the magnitude of the mixed error using Equation 1.15 as long as,

$$\Delta T = \overline{T} - T_0,\ \Delta Q = \overline{Q} - Q_0,\ \text{and}\ \Delta Y = \overline{Y} - Y_0.$$

Equation 1.12 is also valid even though constant errors are present.

The estimation of mixed errors (size of the error interval) could get very involved. The most direct method to compute the error interval for T is to take the magnitude of the differential times the indeterminant error for each term in Equation 1.15, namely

$$|\Delta T| = |(\partial f/\partial Q)(\Delta Q)| + |(\partial f/\partial Y)(\Delta Y)| \tag{1.19}$$

where $|\Delta T|$ is the size of the error interval for T, with $\Delta T$ the indeterminant error in T (e.g., Wall et al., 1972). This equation always yields the maximum value for the error of the quantity to be determined, and is the basis for several different rules concerning the propagation of indeterminate errors (mixed errors) by use of error intervals. Three of the most useful appear to be:

1. If the result T is the sum or difference of two measured quantities Q and Y, the indeterminate error in T is the sum of the errors in Q and Y. Ex. $\text{Mass}_{total} = \text{Mass}_{dry} + \text{Mass}_{vapor}$
2. If the result T is the product or quotient of two measured quantities Q and Y, the percentage error in T is the sum of the percentage errors in Q and Y. Ex. Mass = Density/Volume, % error in M = $\Delta M/M$ times 100.

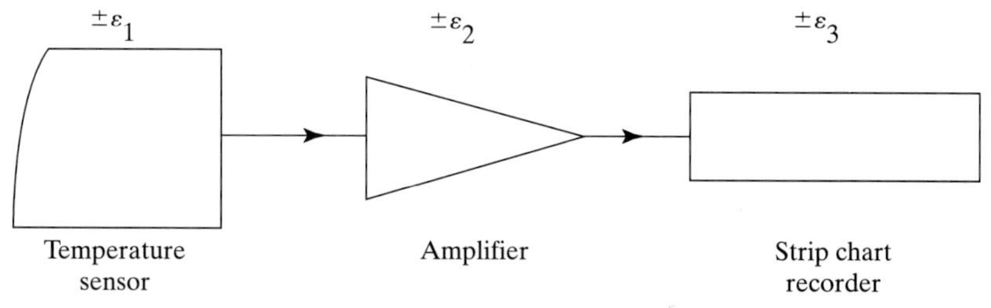


**Fig. 1.6** *A schematic for a simple temperature measuring and recording device. The epsilons denote the errors of the particular device below them.*

3. If the result T is some power $n$ of the measured quantity Q, then the percentage error in T is $n$ times the percentage error in Q. Ex. Volume of Sphere = 4.19 times $R^3$.

**Significant Figures** When recording data and results, or values derived from data it is customary to keep only those figures that are trustworthy and have some significance, otherwise known as *significant figures*. Examples of how to determine the number of significant figures that should be reported follow:

- 66.928 ± 0.001 gm has 5 significant figures; the last one, 8, is in doubt by 1 unit as indicated by the amount of error. 0.246 ± 0.002 gm has 3 significant figures; the last one, 6, is in doubt by 2 units. The first 0 does not count as a significant figure because it is put in simply to emphasize the position of the decimal point.
- 8.306 ± 0.01 is rounded to 8.31 when reporting it. (If the unwanted digit is ≥0.5 then round up; else round down).
- If the error in the value of a quantity is 1 part in 100 (1%), the number of significant figures in this value will never be more than three, although it may sometimes be two. The 1% error of 5.024, 1.135, and 9.807 (to one place) is 0.05, 0.01, and 0.1 respectively. Thus the values with their errors may be written 5.02 ± 0.05, 1.14 ± 0.01, 9.8 ± 0.1. Note that the first two values have three significant figures, while the last one has two.
- Errors of 0.01%, 0.1%, and 1% are used for angles expressed to the nearest 1, 6, and 30 minutes, respectively.
- It is also worthwhile to note that the use of digital computers may introduce further errors, which certainly cannot result in more significant figures then the recorded data.

In summary, the basic principles for establishing a measurement program and for ensuring its success have been laid out in this chapter. Now all one needs is a problem and a little bit of practice, and there is an exercise that will start you out. There are still some more basics to discuss, but this will be most efficiently done with respect to the possible kinds of meteorological measurements that we may need to make in order to address a problem.

## *Exercises*

1. Define the following terms.
   a. Hysteresis
   b. Systematic error
   c. Random error
   d. True value
   e. Significant figures
   f. Ohm's law
   g. Flow of current
   h. Series and parallel circuit
   i. Wheatstone bridge

2. What is the error in the temperature measured by the device shown in Fig. 1.6? Assume the following input/output expressions for the given components:

   component 1: $V_{out} = a + (bT)$, component 2: $V_{out} = cV_{input}$,
   component 3: $X = dV_{input}$, common expression: $X = d\{c[a + bT]\}$.

3. The number of activated nuclei, N, is known to be a function of supersaturation, s, only. Find the most probable error in s, given the following functional form of N with respect to s; $N = Cs^k$, s is in percent; and $C = 400\ cm^{-3}$, $k = 1.0$, the error in $N = \pm 10\ cm^{-3}$, and $s = 0.92\%$.

4. Use the log derivative method to obtain an expression for the maximum error in R. Given that $R = V(I)^{-1}$.

5. There are three common rules (based on $|\Delta T| = \left|\frac{\partial f}{\partial Q}(\Delta Q)\right| + \left|\frac{\partial f}{\partial Y}(\Delta Y)\right|$ where $|\Delta T|$ is the size of the error interval for T) to estimate the error in a result, T. Prove these three rules to be valid using the following examples.
   a. Assume the result is the difference between two measured quantities, Q and Y ($T = Q - Y$). Let Q = mass of a bulb with air = 30.612 g; $\Delta Q = \pm 0.001$ g. Y = mass of bulb without air = 30.601 g; $\Delta Y = \pm 0.001$ g. T = 0.011 ($\pm 0.002$) g.
   b. Assume the result is the quotient of two measured values, Q and Y. $T = QY^{-1}$. Let Q = 150.8 g; $\Delta Q = \pm 0.1$ g. Y = 29.12 $cm^3$; $\Delta Y = \pm 0.05\ cm^3$. T = 5.18 ($\pm 0.01$) g $cm^{-3}$.
   c. Assume that the result is a power law, n, of a measured quantity, Q. Namely, $T = (\pi Q^3)/6$. Where Q = 9.15 $\mu m$; $\Delta Q = \pm 0.03\ \mu m$; T = 401 $\mu m^3$; Also, identify the error in T.

6. What is the equivalent resistance of the following resistors in (A) series and (B) parallel. Resistors are: (1) 680 Ω, (2) 410 kΩ, (3) 3300 Ω, (4) 220 kΩ, (5) 1 MΩ.

7. Given the circuit in Fig. 1.7, Find $I_1$, $I_2$ and $R_{equivalent}$. Use V = 1.5 volts, $R_1 = 150\ \Omega$, $R_2 = 300\ \Omega$.

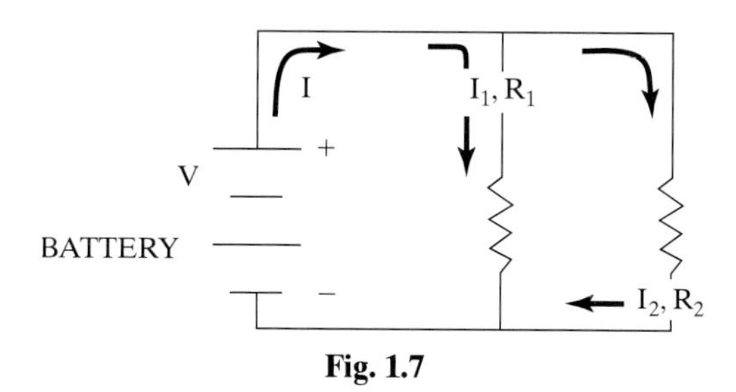


**Fig. 1.7**

8. Given the circuit in Fig. 1.8, Find $I_1$, $I_2$ and $R_{equivalent}$. Use V = 4.0 volts, $R_1 = 400\ \Omega$,

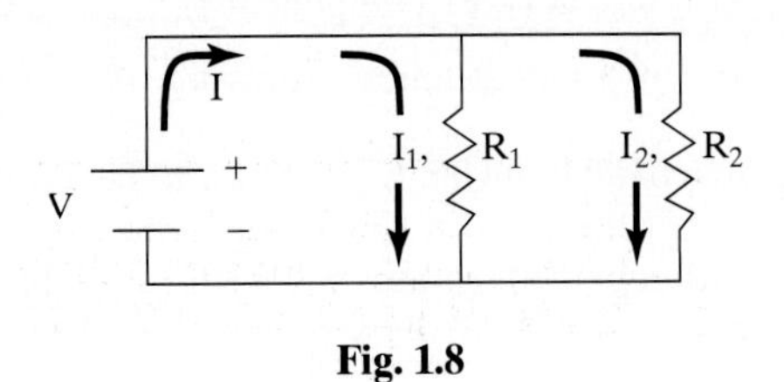


**Fig. 1.8**

$R_2 = 800\ \Omega$.

9. Find the power loss in each resistor element in the circuit in Exercise 8. Also determine the voltage at points A and B.

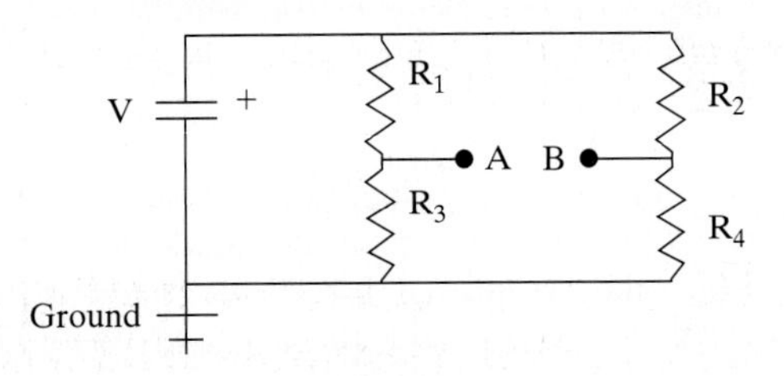


**Fig. 1.9**

10. What is the equivalent resistance of the circuit shown in Fig. 1.9?

11. Calculate the digital number derived from a bipolar 10 bit binary ADC if the following voltages are applied to the input. Let the reference voltage equal 5 V. See Appendix B.
    a. −4976.4 mV
    b. −188.6 mV
    c. 370.9 mV
    d. 3621.3 mV

12. Prepare a QA plan for determining the time of day (to the nearest second) at which the air temperature reaches its maximum value.

# CHAPTER TWO

# Radiation Measurement

The fundamental principles of radiation and its measurement are introduced in this chapter. In particular, we will learn about the instruments that measure the radiation from the sun, earth and its atmosphere, via stationary and mobile platforms. These instruments include radiometers, radars, and lidars.

## 2.1. *Radiation Principles*

Radiation is the process by which thermal energy is transferred from one point to another in the absence of an intervening medium. The radiant energy from the sun drives atmospheric motions that in turn may produce precipitation and other physical phenomenon. The radiation emitted by the sun varies in wavelengths from about $10^{-14}$ to $10^{10}$ meters (m), and the entire range of wavelengths is called the *electromagnetic spectrum.* Waves at the short end are comparable in size to an atomic nuclei, while waves at the long end are about 0.1 times the distance between the earth and the sun. Radiation with wavelengths from less than $\approx 3 \times 10^{-7}$ to $\approx 4 \times 10^{-6}$ m is considered shortwave radiation, and radiation with wavelengths from $\approx 4 \times 10^{-6}$ to $\approx 5 \times 10^{-5}$ m is considered longwave radiation. Visible radiation has a wavelength range between $\approx 4 \times 10^{-7}$ to $\approx 7 \times 10^{-7}$ m. The longwave radiation primarily originates from the earth's surface or in the atmosphere and is also called terrestrial radiation. The shortwave radiation primarily originates from the sun.

Radiant energy is either going toward (emitted, or irradiated), or coming from (radiated) a direction, and is commonly expressed in terms of a flux (or the amount of radiant energy passing an area per unit time), namely, $dQ\,(dt)^{-1}$, Q = energy/area. The radiant energy per unit time moving toward a specific direction and passing through a unit area perpendicular to that direction is called irradiance, E, (also termed radiant flux density, or radiant *excitance*) and it is represented by

$$E = d^2Q\,(dA\,dt)^{-1}. \tag{2.1}$$

The units of E are watts (W) per meter (m) squared, W $m^{-2}$. [Note: 1 watt (W) = 1 Joule per second $\approx$ 697.6 W $m^{-2}$]. The radiant energy per unit time coming from a specific direction and passing through a unit area perpendicular to that direction is called radiance, L. The radiance is also referred to as intensity. The radiance and irradiance are related by:

$$L = dE(\cos\theta\, d\omega)^{-1} \quad \text{or} \quad E = \int_0^{2\pi} L \cos\theta\, d\omega \tag{2.2}$$

where, $d\omega$ is the differential element of the solid angle equal to $\sin\theta\, d\theta\, d\phi$, $\phi$ is the azimuth angle, $\theta$ is the angle between the "beam" of radiation and the direction normal to the horizontal surface on which the radiance is measured (Fig. 2.1). Radiance is not defined if the radiant energy propagates in a single direction (termed parallel beam radiation) since $d\omega = 0$ in this case. Radiation that is uniform in all $d\omega \neq 0$ is called isotropic. Given isotropic radiation, then Equation 2.2 may be simplified to

$$\begin{aligned} E &= \int_0^{2\pi} L \cos\theta\, d\omega \\ &= \int_0^{2\pi} L \cos\theta\, 2\pi \sin\theta\, d\theta \\ &= \pi L. \end{aligned} \tag{2.3}$$

Non-isotropic radiation cases require numerical techniques and have more complicated expressions for E. It is useful to define E and L as functions of wavelength, $\lambda$, since (i) absorptive or emissive properties of materials are a function of wavelength, (ii) the sun's radiation is distributed over various wavelengths, and (iii) radiation instrumentation obtain measurements within specific wavelength intervals or bands. Thus,

$$E = \int_0^{\infty} E_\lambda\, d\lambda \quad \text{and} \quad L = \int_0^{\infty} L_\lambda\, d\lambda \tag{2.4}$$

where $E_\lambda$ is monochromatic irradiance, and $L_\lambda$ is monochromatic or spectral radiance. Note that the wavelength is related to the frequency of the radiation by $\nu = c\lambda^{-1}$, where c is the speed of light.

The radiance from a blackbody is approximated by Planck's law using:

$$L_\lambda(T_{bb}) = (2hc^2\lambda^{-5})\{e^{[(hc)(k\lambda T_{bb})^{-1}]} - 1\}^{-1} \tag{2.5}$$

where h is Planck's constant, k is Boltzmann's constant, $T_{bb}$ is defined as the blackbody temperature of the object, and c is the speed of light. The sun, for example, radiates as a blackbody with $T_{bb} = 6000°K$.

The integration of Planck's law with respect to wavelength yields the radiant excitance, E, from a blackbody and is known as the Stefan-Boltzmann law:

$$E = \int_0^{\infty} \pi L_\lambda(T_{bb})\, d\lambda = \sigma_{sb} T_{bb}^4, \tag{2.6}$$

where $\sigma_{sb}$ is the Stefan-Boltzmann constant = $5.67 \times 10^{-8}$ W m$^{-2}$K$^{-4}$. Recall that since a *blackbody* is defined as an object that emits all radiation that is incident upon it, or the object's emittance equals 1, then Kirchoff's law states that this object

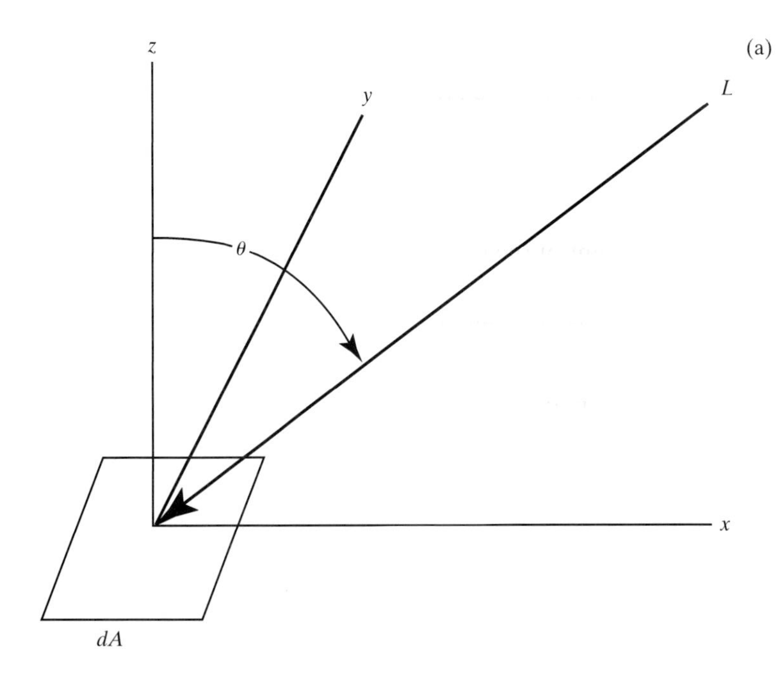


**Fig. 2.1** *The interrelationship between radiance, L, the angle, θ, between the "beam" of radiation and the direction normal to a horizontal surface, and the solid angle, dω, impinging: (a) a horizontal plane surface of differential area, dA (adapted from Fleagle and Businger, 1980), and (b) The "equatorial plane" of a sphere with radius, r. (Adapted from Wallace and Hobbs.)*

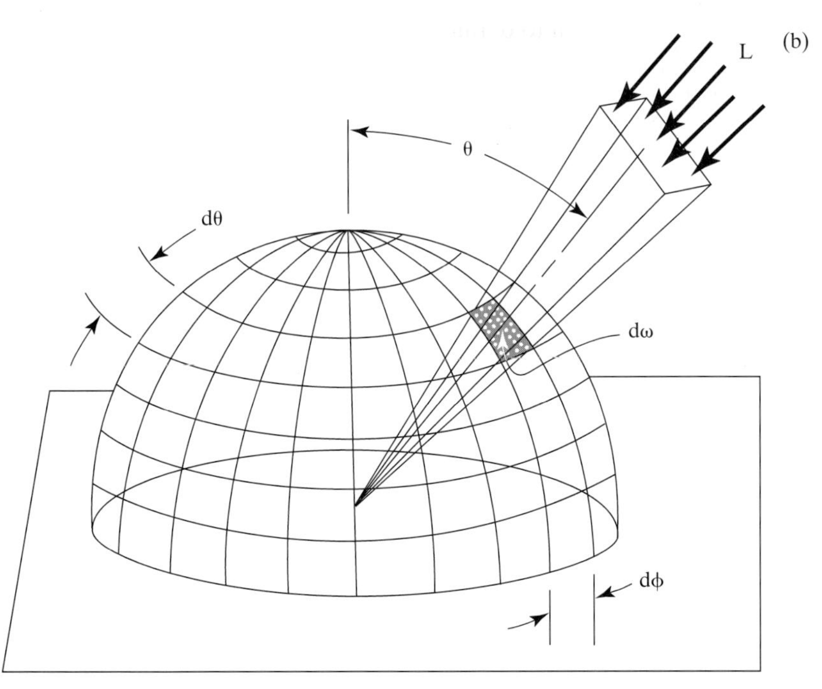

absorbs as well as it emits independent of wavelength. The radiation emitted by the sun is neither mono-chromatic nor isotropic. The sun is considered a blackbody since its emissivity =1 independent of wavelength. The general form of Kirchoff's Law is a function of temperature and wavelength

$$\varepsilon_{\lambda,T} = a_{\lambda,T} \tag{2.7}$$

where $\varepsilon$ and a are the emissivity and absorptivity, or the ratios of actual emitted and absorbed radiation to the maximum possible emitted and absorbed radiation at that temperature. If the blackbody radiates equally in all directions, then E for a blackbody is $= \pi L_\lambda(T_{bb})$. The earth may also be considered a blackbody.

The wavelength of maximum radiance for a blackbody is found by partial differentiating Planck's law with respect to wavelength, setting that expression equal to zero and solving for $\lambda$. This yields the Wien's displacement law:

$$\lambda_{max.} = 2897.8\ \mu m\ ^\circ K T_{bb}^{-1}. \tag{2.8}$$

---

**Example 2.1**

Calculate the $\lambda_{max.}$ for a body radiating at T = 6,000 °K and at T = 288 °K. $\lambda_{max.} = 0.5$ $\mu$m for the object with temperature 6,000 °K and $\lambda_{max.} = 10.1$ $\mu$m for the object with temperature 288 °K.

---

Please see Kondratyev (1969), Liou (1980) and Goody and Yung (1989), for example, if you desire further details on radiation theory.

## 2.2. *Measurement of Solar Radiation*

The measurement of radiation within the earth-atmosphere system is conducted by either measuring all radiation from all directions (i.e., *global radiation*), the component of radiation that is coming directly from the radiating object (i.e., *direct radiation*), or the component of radiation that has been scattered or diffusely reflected (i.e., *diffuse radiation*). For example, global solar radiation is the total flux of solar radiation received from the solid angle of the sun's disc (known as direct solar radiation) plus radiation that has been scattered or diffusely reflected in traversing a horizontal surface of the atmosphere (known as diffuse solar radiation). Diffuse solar radiation is defined as a non-isotropic flux of solar radiation from the sky incident on a horizontal surface and approaches a maximum as one nears the sun. Diffuse radiation is expected to be higher on a day with cumulus clouds and relatively lower on a day with clear skies. Total radiation is defined as shortwave plus longwave radiation, and net radiation is defined as total downward radiation fluxes minus total upward radiation fluxes. During the daytime, global solar radiation can be subtracted from total hemispheric radiation to obtain longwave hemispherical radiation.

Our radiation measuring device will record the electronic response or a change in the surface temperature of a small receiving surface that results from a change in radiation input. Some devices measure a chemical response to a radiation input. Pyrradiometers (or radiometers), pyranometers, pyrheliometers, and pyrgeometers are the four basic kinds of instrumentation for measuring radiation, and their generic use is summarized in Table 2.1. Figure 2.2, Fig. 2.3, and Fig. 2.4 show a net radiometer, pyranometer and a Climatronics Corp. Model CM3 pyranometer, respectively. The CM3 is a modern-day version of the Eppley black and white pyranometer, (e.g., see Fritschen and Gay, 1979). The measuring devices highlighted in Table 2.1 are primarily thermal sensitive devices and they have similar principles of

**Table 2.1** *Examples of basic radiation measuring devices and their application*

| *Device* | *Application*[a] |
|---|---|
| Radiometer | Long and short wave radiation |
| Pyranometer[b] | Global solar radiation over $0.3 \leq \lambda \leq 5.0\ \mu m$. This instrument may be inverted to obtain direct beam albedo, or the albedo in the infrared or ultraviolet range. |
| Pyrheliometer[c] | Direct beam solar radiation usually $0.3 \leq \lambda \leq 4.0\ \mu m$. |
| Pyrgeometer | Longwave atmospheric radiation of $\lambda \geq 4.0\ \mu m$. |

[a] $\lambda$ is wavelength.

[b] Also called solarimeter.

[c] Note that the word pyrheliometer has been associated with global solar radiation.

(a)

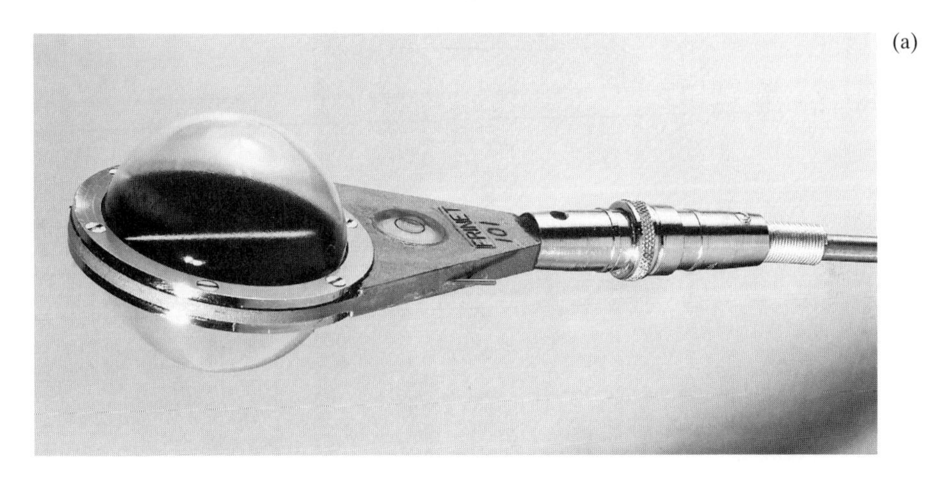


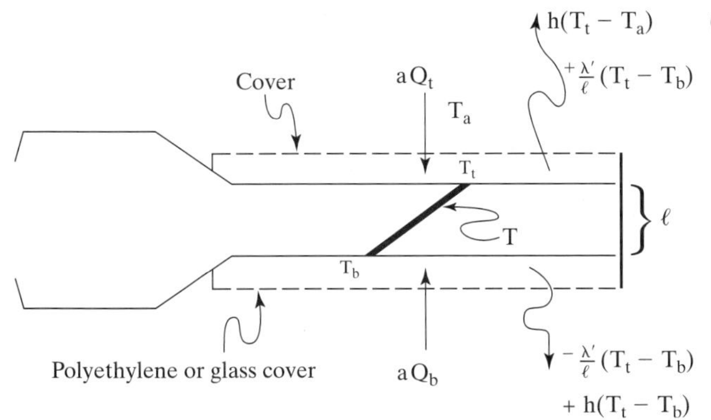


**Fig. 2.2** *(a) A photo of a net radiometer. (Courtesy of D. Gilmore and D. Katz of Climatronics Corp.) (b) Radiative balance of the net radiometer.*

**Fig. 2.3** *A photo of a pyranometer. (Courtesy of D. Gilmore and D. Katz of Climatronics Corp.)*

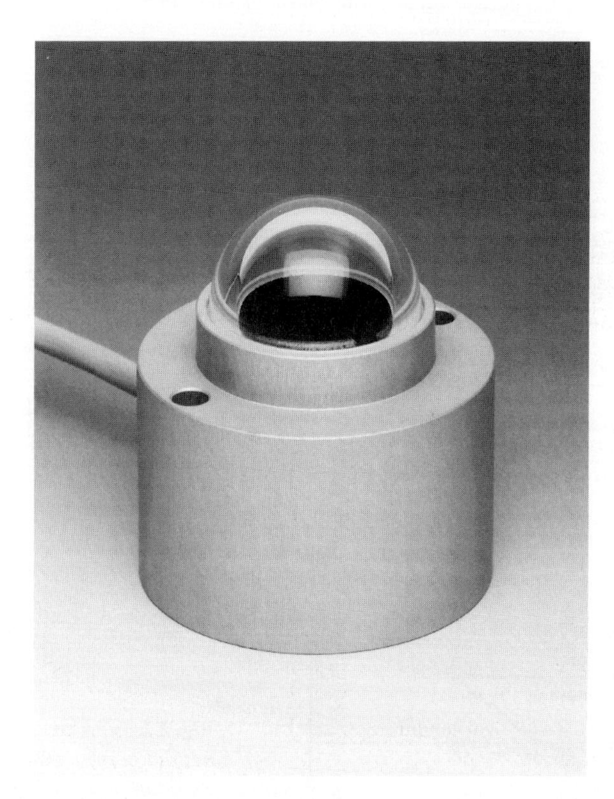

**Fig. 2.4** *A photo of the Climatronics CM3. The CM3 is a rugged, accurate, stable, and low-cost thermopile-type pyranometer that measures diffuse and direct solar radiation in the wavelength range of 0.3 to 3.0 $\mu m$. Its millivolt output is directly proportional to the amount of insolation. (Courtesy of D. Gilmore and D. Katz of Climatronics Corp.)*

operation. Consequently, the operation of a net radiometer will be presented below, since it best describes the operation for all of these devices.

Keep in mind though that the radiation measuring devices today, while similar in principle, have improved some since the early 1980s. See Table 2.2, for example. Table 2.2 compares the specifications of some of the present day commercially available radiation instrumentation with those highlighted in Brock (1984). The accuracy for radiometers depends on a number of factors (Brock, 1984) and is not often specified by the manufacturer. The factors include the minimum radiant energy difference that yields a measurable output (sensitivity), the long term change in calibration (stability), variation of a response as a function of ambient temperature (temperature effect), constancy of response at all specified wavelengths (spectral selectivity), proportionality constant between incident radiation and sensor response (linearity), time constant (i.e., 63.2% response to step change in incident radiation), the deviation of the directional response from that assumed

**Table 2.2** *Accuracy specification of some commercially available radiation sensors*[a]

| *Sensor* | *A* ($\pm mW\ cm^{-2}$) | *B* ($\pm$%) | *C* ($\pm$%) | *D* ($\pm$%) | *E* (s) | *F* (%) |
|---|---|---|---|---|---|---|
| Net Radiometer ($\lambda$: $\approx 0.3-\approx 50\ \mu m$) | 0.1 | 1 | 0.1 | 2 | 10.5 | 1[f] |
| UV Radiometer (Selenium photoelectric cell) ($\lambda$: 0.295–0.385 $\mu m$) | 0.1 | 1 | 0.1 | 2 | 0.01 | 2.5 |
| SUV-100 SpectroRadiometer[b] ($\lambda$: 0.28–0.62 $\mu m$) | 0.02 | 0.5 | 0.2 | 0.1 | 0.1–10 | — |
| Photovoltaic Pyranometer[c] ($\lambda$: 0.4–1.1 $\mu m$) | 0.1 | <2 | 0.15 | <1 | 0.00001 | — |
| UV Photovoltaic Pyranometer[d] ($\lambda$: 0.4–0.7 $\mu m$) | 0.1 | 2 | 0.15 | 1.0 | <0.001 | — |
| Silicon Cell Pyranometer (dome)[d] ($\lambda$: 0.35–1.15 $\mu m$) | 0.1 | 2 | 0.15 | 1.0 | <0.001 | — |
| Precision Spectral Pyranometer ($\lambda$: 0.285–2.8 $\mu m$) | 0.1 | 1 | 0.1 | 0.5 | <0.001 | — |
| Reference Standard Pyrheliometer[e] | 0.2 | 0.2 | 0.2 | 0.5 | 25 | — |
| 1st class Pyranometer[e] | 0.1 | 1 | 1 | 1 | 25 | 3 |
| 2nd class Pyranometer[e] | 0.5 | 2 | 2 | 2 | 60 | 5–7 |
| 1st class net Pyrradiometer[e] | 0.1 | 1 | 1 | 1 | 30 | 5 |

[a]Commercially available sensors as of 1996. A = sensitivity, B = stability, C = temperature effect, D = linearity, E = time constant, F = cosine response. Azimuthal response, spectral sensitivity were <±1% for the pyranometers or otherwise unspecified.

[b]Accuracy for $\lambda = \pm 0.01$ nm. Optional fiber optic coupler estimated to lower minimum signal level reported under accuracy by 20–50 times. Value for stability is over 2 months.

[c]Value for stability is over 1 year; ±5% accuracy on readings.

[d]Same as c, except ±5% accuracy under natural daylight conditions.

[e]Derived from Table 4-6 of Brock (1984). Note: in Brock (1984) class is related to standards (See glossary).

[f]The values shown in this Table for the non-Brock devices are from 0 to 70° zenith angle. The net radiometer has a cosine response ±3% between 70–80°. All present commercially available instruments are cosine corrected up through 70 or 80°.

(cosine response), the deviation of the response due to azimuthal angle of orientation of the instrument (azimuthal response), and, for pyrheliometers, the amount of circumsolar radiation allowed by the instrument's field of view (aperture). An additional factor is the accuracy of the recording device. It should be noted that an instantaneous value of solar radiation is not necessary for most meteorological applications, and a 2 minute average or longer is relevant. Nonetheless, please take advantage of the nearly instantaneous time response, somehow, especially in modelling efforts, and, of course, research.

*Radiometers (Pyrradiometers).* The thermal sensitive radiation devices are the most common. Their principle of operation, whether they utilize resistance elements or thermopiles, is described using a net radiometer (Fig. 2.2b), or a net pyrradiometer. The following is primarily based on Fritschen and Gay (1979). In general, radiometers produce a signal due to a differential absorption of radiation. The differential absorption is achieved by detecting the temperature differences of two surfaces that are exposed to radiation, such as that between a black and white surface, two black surfaces, or a blackened surface and the base or housing of the instrument. Since the principle of operation is independent of the absorbing surfaces, then let's place a blackened plate containing a thermopile (or thermal transducer—see Chapter 3) into a radiation field because its response is independent of wavelength. The radiation field has a radiant flux density of $Q_t$, and our plate has a thickness $\ell$. The plate comes into equilibrium fairly quickly, and its energy budget may be expressed for the top of the plate by Equation 2.9:

$$a Q_t = [\varepsilon \sigma T_t^4] + [h(T_t - T_a)] + [(\lambda' \ell^{-1})(T_t - T_b)] \qquad (2.9)$$

where h is the thermal convection coefficient, $\lambda'$ is the thermal conductivity, $T_a$ is the temperature of the air, and $T_b$ is the temperature of the bottom of the plate. Similarly, the bottom plate energy balance may be expressed by Equation 2.10:

$$a Q_b = \varepsilon \sigma T_b^4 + [h(T_b - T_a)] - [(\lambda' \ell^{-1})(T_t - T_b)] \qquad (2.10)$$

The Net Radiation, $Q^*$, = (2.9) − (2.10)

$$Q^* = a(Q_t - Q_b) = [\varepsilon \sigma (T_t^4 - T_b^4)] + [h(T_t - T_b)] + [2(\lambda' \ell^{-1})(T_t - T_b)]$$
(2.11)

If cover was glass, for example, instead of polyethylene, then the h term could be ignored, and by using the approximation that $(T_t^4 - T_b^4) \approx 4T_t^3(T_t - T_b)$, Equation 2.11 may be rewritten as

$$Q^* = a(Q_t - Q_b) = [(4\varepsilon \sigma T_t^3)(T_t - T_b)] + [(2(\lambda' \ell^{-1}))(T_t - T_b)]. \qquad (2.12)$$

Furthermore, if a thermoelectric potential, V, is set up as a result of the temperature difference, then $V = NC(T_t - T_b)$ where C is a constant of proportionality and N is the thermoelectric power, then

$$Q^* = \{(4\varepsilon \sigma T_t^3) + [2(\lambda' \ell^{-1})]\}[V(NC)^{-1}] = gV, \qquad (2.13)$$

where g is the calibration constant that depends on the temperature, emissivity, conductivity, and the thickness of the thermopile. [Note that the first term in the fancy brackets is small compared to the second term and could be neglected]. The use of polyethylene instead of glass keeps the h term in Equation 2.11, but it allows a broader $\lambda$ range to reach the sensor. The glass limits the amount of longwave radiation reaching the sensor.

Radiation measurement devices should ideally have sensors that are good absorbers over the specified wavelength band, and at the same time have a good cosine response. Table 2.2 indicates that the sensors have improved cosine responses from those 15–20 years ago. The time constant, sensitivity problem mentioned by Fritschen and Gay (1979) also appears to be less significant.

*Angstrom Compensation Pyrheliometer.* The Angstrom Compensation Pyrheliometer has been used since 1890, invented by K. Angstrom of Sweden, and is one of the best known and most reliable instruments for measuring solar intensity. Consequently, it is now used as a calibration standard and it serves as an example for this category of radiation measuring devices. It is the standard pyrheliometer referred to in Table 2.2. The sun's rays are directed into its aperture at an incidence angle perpendicular to the surface of two blackened manganin strips ($\approx 20 \times 2 \times 0.01$ mm) mounted at the opposite end from the aperture. One of the strips is exposed to the sun while the other is shaded from it. The shaded strip is then electrically heated. This "heating" current is regulated by thermocouples such that the temperature of the two strips becomes identical. A null-balance of a sensitive galvanometer is employed to achieve the latter. The solar flux density, E, of the solar radiation is related to the "heating" current, i, by the formula

$$E = J i^2, \tag{2.14}$$

where J is the instrument constant, E has units of W $cm^{-2}$, and i has units of amps. J is usually determined by comparison with another standard instrument, but it may be calculated if you know the resistance ($R_{electrical,heat}$), width of the manganin strip ($b_{strip}$), the absorption of the surface of the strips ($a_{strip}$), and the electrical heat equivalent:

$$J = R_{electrical,heat} (a_{strip} b_{strip})^{-1}. \tag{2.15}$$

The major advantage of this instrument is that it uses the principle of temperature compensation to yield an increased accuracy in making individual measurements. Its high temporal stability, fast response, high sensitivity, and ease of operation add to its advantages. However, as with most radiometers used in absolute measurements, the drawbacks include: an incomplete elimination of traces of sunlight entering the aperture; the imperfect simulation of a blackbody with respect to absorption and uniformity over the entire sensing surface, as well as the difficulty in constructing two manganin strips that are identical in shape and thermal capacity. The latter may be alleviated by making measurements from alternate exposures of each strip. The mean value of several successive measurements is then the output signal. Pyrheliometers must always be aimed at the sun, especially when measuring the direct normal radiation, and, consequently, the sun tracker must also be maintained.

Radiation measurements are made according to the absolute radiation scale otherwise known as the World Radiometric Reference (Brock, 1984). Periodic calibration of the aforementioned instrumentation is a must in order to reduce any potential longterm stability induced errors. One should try to use reference standards, such as the Angstrom Compensation Pyrheliometer. Pyranometers have been calibrated with incandescent lamps, although one may also calibrate a pyranometer against an Angstrom Compensation Pyrheliometer using a shading/non-shading method while in the field under clear skies. Booth et al. (1995) describe a comprehensive calibration scheme for a number of radiation instruments, including a spectral radiometer. Note that using a variation of the shading/non-shading method it is possible to obtain the diffuse radiation from a pyranometer and pyrheliometer, namely

$$\text{Diffuse Radiation} = (G) - (S \cos \Theta_0) \tag{2.16}$$

where G is the global radiation measured by the pyranometer, S is the direct normal radiation from the pyrheliometer, and $\Theta_0$ is the average solar zenith angle during the measurement.

NASA plans to study the sun, the solar wind, and other related phenomena in the near future. One of the instruments that will be used is called a sun photometer (Matsumoto et al., 1987). The platform for this instrument is presently airborne, although it has been used on the ground. It should be mentioned that the sun photometer may also be able to provide information about cirrus clouds and aerosols.

## 2.3 *Measurement of Radiation Using Remote Sensing Devices for Weather*

Remote sensing refers to the gathering of information about the earth and its environment from a distance. Examples of remote sensing measuring devices include: radar, lidar, sonar, radiometers, combinations of the latter, and a few that are presently being developed, such as the lidar system in space, or LITE (McCormick et al., 1993). These devices can be deployed on ground-based, airborne, or satellite sampling platforms from which they can retrieve spatial information on windspeed and wind direction, radiation, surface and air temperature, and water vapor content, as well as information on aerosols, clouds, ice, and liquid phase precipitation.

*In-situ measuring* refers to the gathering of information from within or on a media, i.e., they are made in the actual location or environment of the object or entity being measured. In-situ measurements are slowly being replaced by remote sensing measurements. Nonetheless, in-situ sampling platforms are still the most common way to make measurements of atmospheric air and soil temperature, atmospheric pressure, windspeed and wind direction, radiation, moisture content, aerosols, clouds, ice, and water phase precipitation.

The purpose of this part of the chapter is not to define and discuss the basics behind each and every measurement that can be made via in-situ or remote platforms, because, as the examples suggest, most measurements of interest to atmospheric scientists and environmental scientists may be obtained either way. Instead the choice is to provide a brief background behind the basic remote sensing devices, such as radars, lidars, radiometers, and satellite sensors. The reader is strongly encouraged to read the plentiful texts and literature that exist for each of the remote sensing devices available today if additional details are desired.

***2.3.1. Radar.*** The sections under this heading are covered in depth by other authors. For example, Battan (1973) authored a standard text on the subject, and accounts of the early history of radar meteorology are given by Atlas (1989). Rogers and Yau (1991) also provide a detailed summary of the weather radar.

A radar measures the range and bearing of backscattering objects or targets. The basic components of the radar are the transmitter/receiver, antenna, and display units (Fig. 2.5). The transmitter generates short pulses of energy with power $P_t$ in the microwave frequency portion of the electromagnetic spectrum. The antenna focuses these pulses into a narrow beam that is directed into a semi-spherical dish. The beam propagates outward at approximately the speed of light, c ($= 3 \times 10^8$ m $s^{-1}$). If the pulses intercept an object (or "target") with different refractive characteristics than air, a current is produced in the object which perturbs the radar pulse and causes some of the radar pulse energy to be scattered. Part of the scattered energy will generally be directed back toward the antenna where it is then redirected to the receiver. The power of the return signal, P, is expressed in terms of the reflectivity factor, and is the so-called Radar Equation for distributed targets (e.g., Battan 1973; Rogers and Yau, 1991):

$$P = C(\text{radar term})(\text{target term}) \tag{2.17a}$$

where

$$C = \pi^3 c(1024 \ln(2))^{-1}, \tag{2.17b}$$

$$(\text{radar term}) = P_t G^2 (\Delta\lambda)^2 t_p \lambda^{-2}, \tag{2.17c}$$

$$(\text{target term}) = |K^2| Z r_r^{-2}, \tag{2.17d}$$

G is the antenna axial gain which is dimensionless,
$(\Delta\lambda)$ is the beam width and is typically 1° for a meteorological radar,
$\lambda$ is the wavelength of the radiation,
$t_p$ is the duration of the pulse,
$|K|^2$ is the squared magnitude of the ratio

$$K = (m^2 - 1)(m^2 + 2)^{-1}, \tag{2.17e}$$

K depends on temperature, wavelength, and the composition of the sphere,
$m = n - ik$ is the complex index of refraction of a sphere
n is the refractive index and k is the absorption coefficient,
$r_r$ is the range of the target relative to the radar,
Z is the reflectivity factor defined as the ratio of the intensity of the return signal compared to that of the initial signal, and is related to the microstructure of the target (i.e., the number densities and sizes of the components of the target)

$$Z = \int_0^\infty N(D) D^6 dD, \tag{2.17f}$$

N(D) dD is the number of scatterers with diameter (D) in the size interval dD per unit volume.

(a)

(b)

(c)

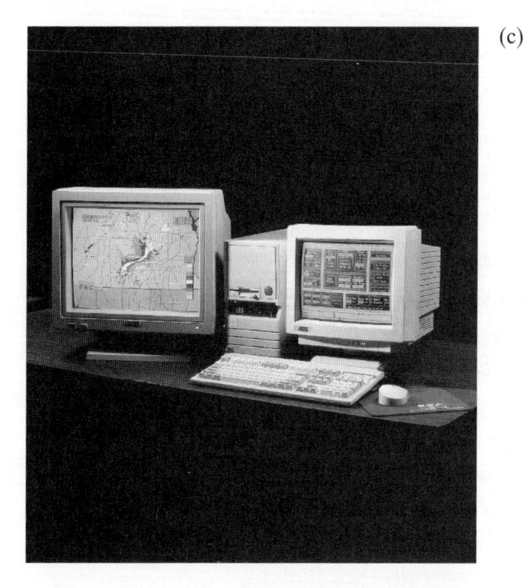

(d)

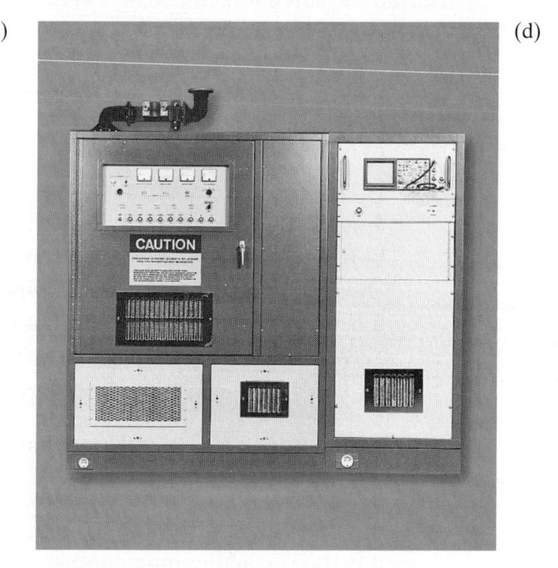


**Fig. 2.5** *(a) A conventional weather radar (Photo courtesy of Tom Henderson, Atmospherics, Incorporated), and its (b) antenna, (c) display, and (d) transmitter/receiver units. (Photos courtesy of H. Quast of Enterprise Electronics Corp.)*

For example, in the case of precipitation, N(D) is the drop-size distribution for raindrops, or the melted diameter distribution for snowflakes. Note that if the latter convention is not used for snowflakes, then the density of the snow would have to appear as a correction to $|K|^2$.

The bearing (direction) of the target is determined by noting the antenna azimuth and elevation at the instant the signal returns from the target. The range, $r_r$, is determined by a timing circuit that counts the time between the transmission and reception of a signal. The transmission frequency of the radar signal defines the maximum range within which targets can be correctly indicated. Since the energy travels at the speed of light, c, the time, t, between transmission and reception is related to the range of the target by $t = 2r_r/c$. The factor of 2 accounts for the round trip to $r_r$ traveled in time t by the signal. The ray paths of the pulses may be assumed to be approximately circular, upward bending paths, or arcs, with a radius of curvature slightly greater than that of the earth (assuming a standard atmosphere). This explains why the radar's horizon is slightly farther than the visual horizon. Abnormal downward bending of the radar waves is an anomalous propagation that occasionally confuses the interpretation of weather echoes or targets.

The clarity of the weather echoes or targets depends on the width of the focused beam sent out by the radar. A narrow beam width will result in a well defined target position, whereas a larger beam width will result in a less clearly defined target. The minimum size of a target that will be detected by the radar depends on the wavelength of the radar signal (Table 2.3). For example, the S band radar is sensitive to precipitation but not cloud, and the K band radar is sensitive to cloud but not precipitation. The standard National Weather Service (NWS) conventional (or non-Doppler) weather radar has a wavelength of 10 cm, while their Doppler weather radar has a wavelength of $\approx$11 cm. The conventional NWS weather radar will sense precipitation sized hydrometeors or larger.

There are at least three ways to display the data received by a radar (Fig. 2.6). The simplest display is the A-scope. The A-scope basically gives the range and target intensity along a fixed azimuth. A more useful Display is the Plan Position Indicator (PPI) Display. The PPI Display is a flat view of the "horizontal" or azimuthal scan of the target(s). The azimuthal scan is generally not a horizontal slice of the target, but actually inclined slice(s) through the target. When comparing radar data with concurrent aircraft data, for example, a horizontal slice through the target is often desirable, and may be accomplished by electronically combining data from different levels. The resulting display is termed a constant altitude plan position indicator (CAPPI) display. The CAPPI is a PPI display scheme that shows a nearly horizontal level radar scan (of the target), and was developed by Marshall in the late 1940s. One final display that is very useful is the Range Height Indicator (RHI) display. This display shows the vertical structure of the target as the radar scans it in the vertical at a fixed azimuth. The vertical structure is important for identifying regions of hail within a cloud, for example.

**Table 2.3** *Radar frequency versus wavelength band versus application*

| *ν(GHz)* | *λ(cm)* | *Band* | *Application* |
|---|---|---|---|
| 37.5 | 0.8 | mm | Cloud (ice), haze |
| 30 | 1 | K | Cloud |
| 10 | 3 | X | Cloud, snow |
| 6 | 5 | C | Onset of precipitation |
| 3 | 10 | S | Precipitation |
| 1.5 | 20 | L | Large precipitation |

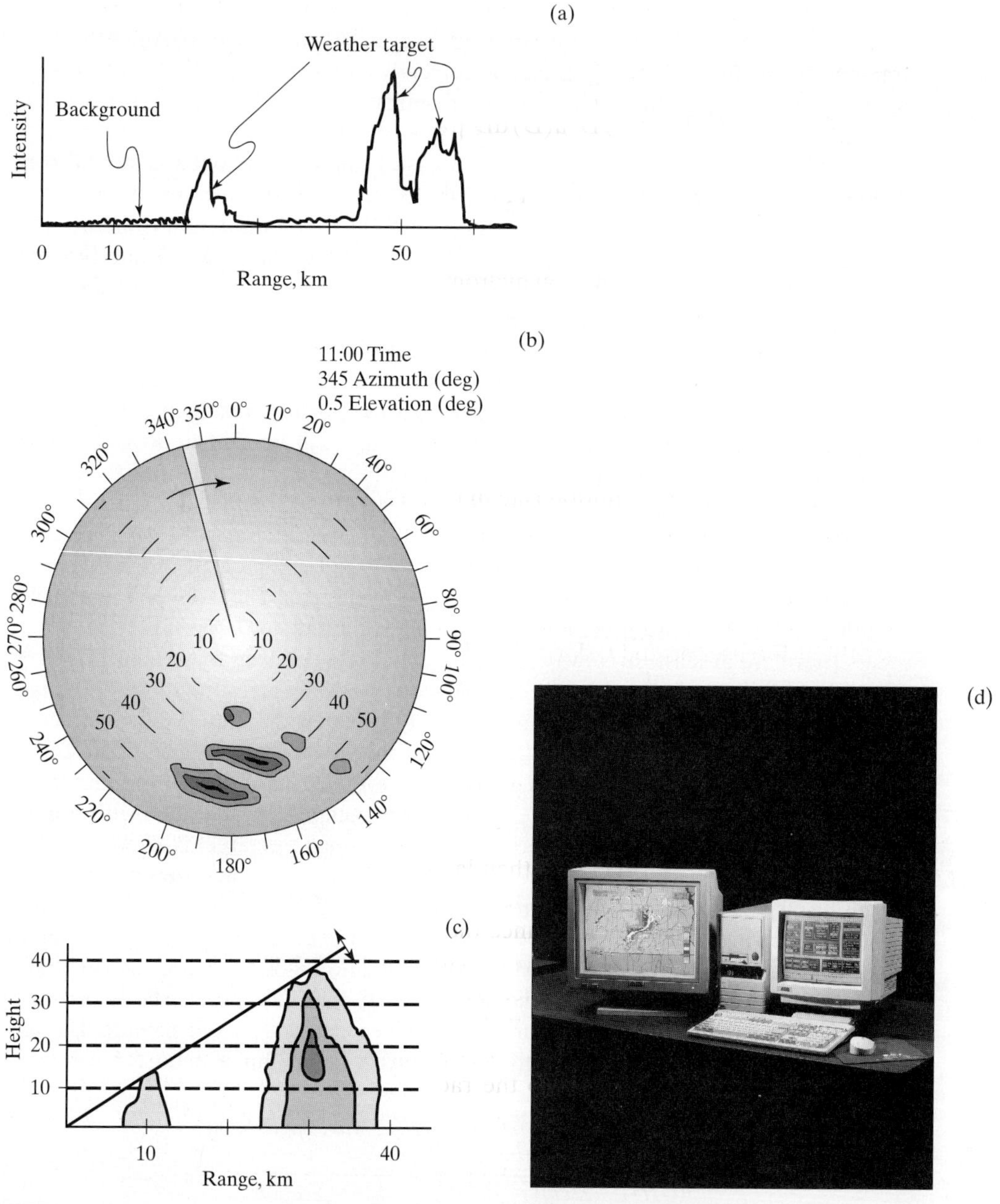


**Fig. 2.6** *Typical displays of the data received by a radar: (a) A-scope, (b) Plan Position Indicator (PPI), (c) Range Height Indicator (RHI) displays, and (d) modern-day display that allows the aforementioned displays to be viewed simultaneously. (Photo courtesy of H. Quast, Enterprise Electronics Corp.)*

The conventional weather radar reflectivity factor, Z, is related to the precipitation rate or intensity, R, expressed in mm $h^{-1}$. The precipitation rate is defined by

$$R \equiv \left[ (\pi/6) \int_0^{\infty} N(D)\, D^3 u(D)\, dD \right], \tag{2.18}$$

where the $\pi/6$ comes from the approximate spherical shape of the precipitation hydrometeors of diameter (D) distributed according to a number density N(D), with fall speeds u(D). The relationship between Z and R is commonly expressed according to the distribution of spherical hydrometeors known as the Marshall–Palmer distribution (Rogers and Yau, 1991, for example) as

$$Z \approx 200\, R^{1.6}, \tag{2.19a}$$

and R is in mm $h^{-1}$. Similarly for snow

$$Z \approx 2{,}000\, R^2, \tag{2.19b}$$

and again R is the precipitation rate in mm of water per hour. In the case of snow, the diameter becomes the melted diameter of the snow (e.g., Rogers and Yau, 1991). The reflectivity is usually expressed in terms of dbZ (decibels relative to reflectivity $= 10 \log Z$), and is routinely grouped into various categories.

Given that the size of the backscatterer is smaller than $\approx 0.1$ times the wavelength of the transmitted radar pulse, then the area of a spherical object doing the backscattering, known as the backscattering cross-section, $\sigma_{bs}$, is given by

$$\sigma_{bs} = 64\, \pi^5 \lambda^{-4} |K|^2 r^6 \tag{2.20}$$

where r is the radius of the spherical backscattering object. Equation 2.20 shows that targets with small K's will return less radiation than targets with larger K's at a given $\lambda$, if the targets are the same size. Smaller-sized targets will return less incident radiation back to the radar than larger targets of the same K, T, at a given $\lambda$. For example, the return signal is less intense for ice than for liquid precipitation hydrometeors of the same size, since the refraction term, K, is approximately 0.21 for ice compared to approximately 0.93 for water over the meteorological range of temperatures. However, melting ice particles (i.e., ice crystals or hailstones with a liquid surface layer) reflect like water creating an increase in radar reflectivity known as the bright band. Smaller-sized precipitation particles will return a smaller amount of the radiation back to the radar than the larger particles of the same phase. Note that the return signal is proportional to the 6th power of the linear hydrometeor size (Equation 2.17f). So if the linear size of a hydrometeor is 2 mm in radius, then it will reflect 64 times the power of 1 mm radius hydrometeors.

The reflectivity from precipitation targets is commonly grouped into 6 levels, namely 1 through 6. These levels are colored differently and often shown on TV weather broadcasts. Levels 5 and 6 are usually observed when hail is present in the cloud. The vertical location of the levels 5 and 6 radar returns relative to the $-5\,°C$ isotherm is also important when determining whether hail will fall from a precipitating cloud.

The precipitation rate, R, may also be expressed in terms of the liquid water content (LWC)

$$R \approx (LWC\,\Delta Z)/(t\rho) \tag{2.21}$$

where $\Delta Z$ is the thickness of the cloud, t is the time for the cloud droplets in $\Delta Z$ to become precipitation-sized, and $\rho$ is the density of water. The quantity (LWC $\Delta Z$) is often referred to as the vertical liquid water path (LWP) of the particular cloud. The LWP is popular in microwave satellite data retrievals of cloud and precipitation properties since the fall speed of the hydrometeor does not have to be known, and the liquid water content is easily determined.

**Doppler Radar** A Doppler radar uses the Doppler shift principle to detect the size and motion of its target. An example of the Doppler shift principle would be the change in noise level as a bus approaches, arrives, and then passes a particular corner. It is common for radars to transmit signals that have the same frequency and the same phase relationship. The Doppler radar has a receiver that essentially compares the frequency of the returned signal to that of the transmitted signal, and any changes in the frequency are interpreted as resulting from a Doppler shift. The frequency shift, $f$, relative to the transmitted signal, is related to the velocity of the target, $V_{target}$, with respect to the radar of wavelength, $\lambda$,

$$f = 2V_{target}\lambda^{-1}. \tag{2.22}$$

In contrast, recall that the conventional radar compares the intensity, or simply amplitude, of the return signal to that of the original transmitted signal.

Doppler radars may either be (i) vertically-pointing, or (ii) horizontally-pointing. The former yields vertical velocities or vertical Doppler speeds of the targets. Since Z is a strong function of target size, and the target size may be used to determine the target's terminal fall speed. Then, the frequency shift that yields Doppler velocity (or target velocity obtained from the frequency shift) represents the actual fall speed in the vertical direction. The horizontally-pointing Doppler radar uses the fact that the target moves with the airflow, and the resulting Doppler speed corresponds to the radial component of air motion to or away from the radar. Doppler radars may be switched to work in conventional reflectivity mode, wherein the amplitude of the return signal is compared to that of the transmitted signal. Note too that conventional radars may be upgraded to also measure Doppler speeds.

Two other kinds of radars are the dual wavelength radar and the polarization diversity radar. The polarization diversity radar transmits radiation of a particular wavelength and polarization, and looks at the polarization and relative phases of the returned signal. It is commonly used to distinguish hail from rain since the reflectivity from conventional radar will not distinguish between phases. Conventional radars are also used on airborne platforms and the interpretation of that data must include accounting for the movement of the radar.

***2.3.2 Lidar.*** Lidar (Fig. 2.7) is the acronym for Light Detection and Ranging, and may be deployed on ground-based, airborne and spaceborne platforms. The lidar is the optical analog to radar, since it transmits radiation in the visible range using a $CO_2$ or $N_2$ laser. The received light energy gets passed through a telescope holding

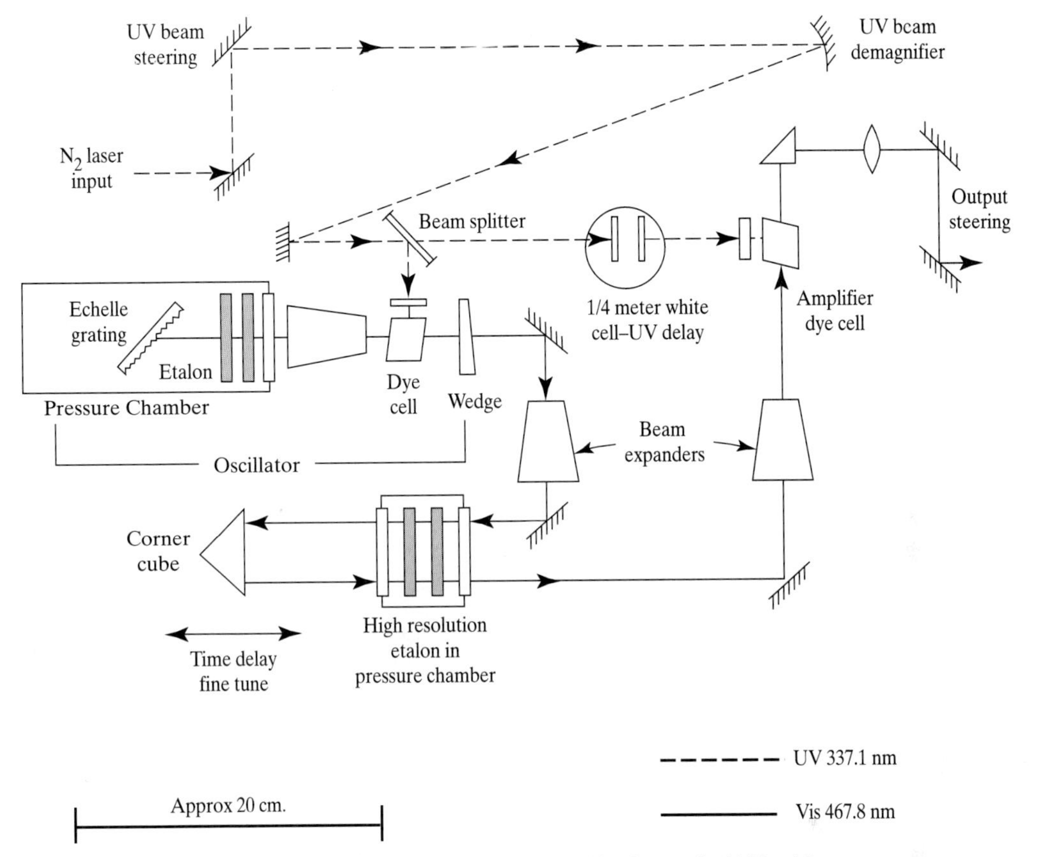


**Fig. 2.7** *A high spectral resolution lidar. (Adapted from Shipley et al., 1983 with permission from American Meteorological Society.)*

receiving optics and electronics to detect the backscattered light energy. It is then amplified and sent into a data acquisition system. Lidar data are continuous, and have vertical and horizontal resolutions within ≈15 and ≈100 m respectively.

The characteristics of the backscattered laser energy, detected at the receiver, provide many details of the composition and dynamics of the atmosphere. Table 2.4 highlights the atmospheric phenomena that the lidar has been used to gather information about. The primary application of the Lidar has been to gather information on aerosols, cirrus clouds, and, most recently, on winds and water vapor profiles. Lidars have been deployed on ground, aircraft, and satellite platforms. Note that the presence of only one reference given under any phenomenon in Table 2.4 does not mean that lidars have not or are no longer used to detect that phenomenon. The references are meant as first steps toward additional information on the use of lidar to study each of the named phenomena. For example, Sassen is still using a lidar to detect precipitation, and Platt for clouds. The depolarization ratio (e.g., Sassen, 1975) is related to the ratio of ice crystal to (supercooled) water droplet concentrations as well as to the cloud thickness. Sassen (1978) gives linear depolarization ratios for backscatter from various types of ice particles: 0.03 to 0.5 for mixed phase clouds, 0.5

**Table 2.4** *The atmospheric phenomena that have been studied by lidars*

| *Atmospheric phenomenon* | *Relevant parameters retrieved* | *Some references* |
|---|---|---|
| Aerosols | extinction<br>optical depth<br>vertical profile | Shipley et al., 1983<br>Sroga et al., 1983<br>McCormick et al., 1993<br>Ellingson and Wiscombe, 1996<br>Chudamani et al., 1996 |
| Clouds (primarily cirrus) | cloudbase height or top<br>cloud extinction<br>optical depth<br>cloud thickness (if cloud ≤100 m thick)<br>cloud cover depolarization ratio | Platt et al., 1994<br>Ellingson and Wiscombe, 1996<br>Spinhirne et al., 1996b |
| Precipitation (especially ice phase) | depolarization ratio<br>extinction<br>optical depth | Sassen, 1977, 1978 |
| Water Vapor | water vapor mixing ratio profile<br>total column water vapor | Goldsmith et al., 1994<br>Ellingson and Wiscombe, 1996 |
| Winds | windfield | Baker et al., 1995 |

for pure ice crystal layers, 0.5 for snowflakes, and 0.6 for rimed ice particles. Smiley et al. (1980) measured a depolarization ratio of 0.2 at Palmer Station in a mixed phase cloud. The depolarization for pure spherical water drop clouds should be zero for single scattering. Multiple scattering causes depolarization in dense water clouds and should not be a problem over the poles. However, a high degree of orientation of hexagonal plate crystals can produce very small values of depolarization which could yield a misinterpretation of there being water present when there is actually none (Smiley et al., 1980). A high degree of orientation means that the ice crystals are not all aligned in the same manner. For example, some have their longest dimension oriented parallel to the ground while others have it perpendicular to the ground and others have it oriented at various angles relative to the ground.

The backscatter signals from the lidar need to be adjusted for instrumental errors, and possibly for (differential) attenuation at each return wavelength depending on the application; ozone and/or other gaseous (e.g. oxygen) absorption; as well as Rayleigh scattering, Mie scattering, and geometric scattering depending on the application (e.g., see references in Table 2.4).

There are a few relatively new lidar developments that warrant mentioning. Shipley et al. (1983) and Sroga et al. (1983) describe a high spectral resolution lidar and its calibration for use as a device to measure optical scattering properties of atmospheric aerosols, while Baker et al. (1995) describe and use a spaceborne lidar wind sensor. Spinhirne et al. (1996) describe an eye safe lidar system (Fig. 2.8), known as the micro pulse lidar (MPL), that has a stable optical alignment, minimal noise induced by background radiation, and a greater than 100:1 depolarization ratio. It is also compact and runs autonomously. It has been used recently to profile atmospheric cloud and aerosol concentrations and has just become commercially available.

A new lidar is being developed for a spaceborne or satellite platform under the lidar in space technology experiment of NASA (McCormick et al., 1993), herein

(a)

**Fig. 2.8** *(a) The micropulse lidar and (b) a schematic of its operation (Courtesy of SESI). (c) A photo of the micropulse lidar mounted in a pod of a NASA ER-2 along with its engineers. (Photo by T. P. DeFelice.)*

(b)

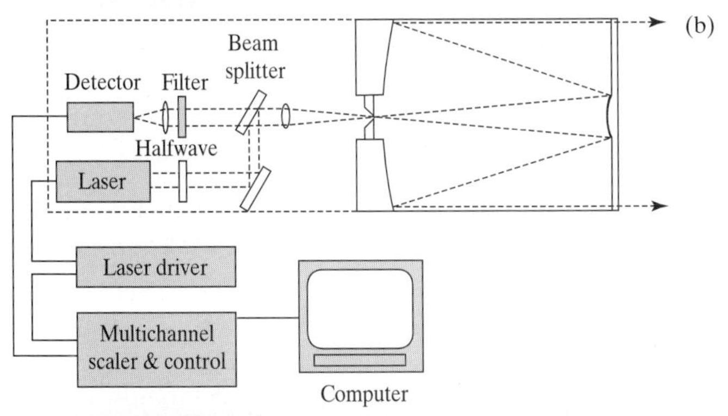


(c)

termed LITE. LITE uses: (a) a three wavelength (1.064 $\mu$m, 0.532 $\mu$m, 0.355 $\mu$m) neodymium, $N_d$:YAG laser and 1 m diameter telescope (Couch et al., 1991); (b) photomultipliers for the submicron channels; and (c) a photodiode for the 1.064 $\mu$m channel to detect the backscattered laser light. The return signals are amplified, digitized, and simultaneously stored and sent to a ground data link. Its field of view will be selectable (i.e., 1.1 mr, 3.5 mr, and opaque) with a sampling interval of 15 m. LITE will be able to derive such spaceborne parameters as stratospheric temperature and density profiles, tropospheric aerosol backscattering characteristics, cloud top and cloud base, optical thickness, number of cloud layers, planetary boundary layer height, and more.

***2.3.3. Radiometer.*** The radiometer is an instrument that measures radiation incident on its aperture. The basic components (Figure 2.9a) of a radiometer consist of: a radiant energy-collecting aperture, a detector, amplifying electronic circuitry, and a device to store the output. The aperture is a lens, mirror, window, or an antenna if it is a microwave radiometer. The detector in an infrared (IR) radiometer converts the impinging radiation into an easily measured quantity such as voltage, or a change in resistance. The microwave radiometer detector may also operate in a similar fashion to the IR detector as long as the converted impinging radiation is amplified to yield a useable signal to noise ratio from the detector without changing its frequency (e.g. Snider and Westwater, 1972). Infrared radiometers are considerably less complicated than microwave radiometers, although the two are operationally similar. Consequently, a brief description of the fundamental characteristics of radiometer operation is made with respect to a microwave radiometer.

The signal received by a microwave radiometer at each of a number of different frequencies (e.g., 19.35, 22, 37, and 85.5 GHz) is compared to a standard through an electronic circuit based on the null-balance wheatstone bridge (e.g., Fig. 2.9b). Simply, the antenna or the sensing element of the radiometer is the unknown resistor that is exposed to the thermal radiation emitted from all radiators between it and the surface of the earth. The resistance of the antenna that balances the flow of current across the bridge circuit is converted to a temperature, known as the antenna temperature, $T_a$. The antenna temperature is essentially the temperature of an equivalent blackbody radiator. There is a linear relationship between $T_a$ and brightness temperature, $T_b$, of a radiating object, known as the Rayleigh-Jeans Approximation that primarily holds in the microwave region of the electromagnetic spectrum.

***2.3.4. Satellite.*** A satellite is simply a platform for measuring devices that commonly include radiometers and other sensors used to detect radiation at specific wavelengths of the electromagnetic spectrum. This portion of the chapter deals with some background on satellite orbits, the scenes viewed by radiometers and the relationship between $T_b$, $T_a$, and radiances from objects between the surface of the earth and the satellite radiometer's aperture.

The orbit of meteorological satellites is ideally circular. An exact expression of the orbit is derived from Newton's law of motion and universal law of gravitation, as summarized by Kepler's law and equation. The actual orbit of a satellite (Fig. 2.10) has an eccentricity, $\varepsilon$, (because the orbit is not usually ideal), with a semi-major axis,

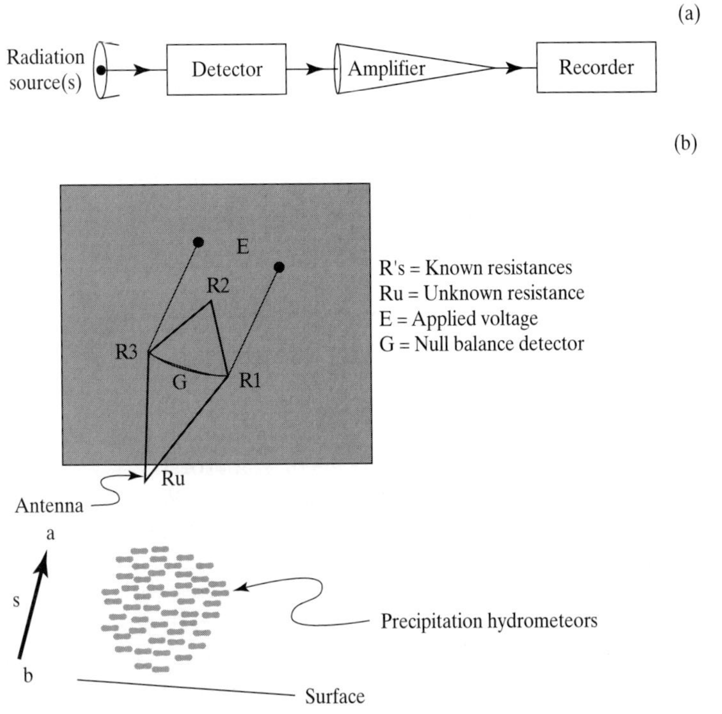


**Fig. 2.9** *(a) The basic components of a radiometer. (b) Conceptual view of a passive spaceborne microwave radiometer.*

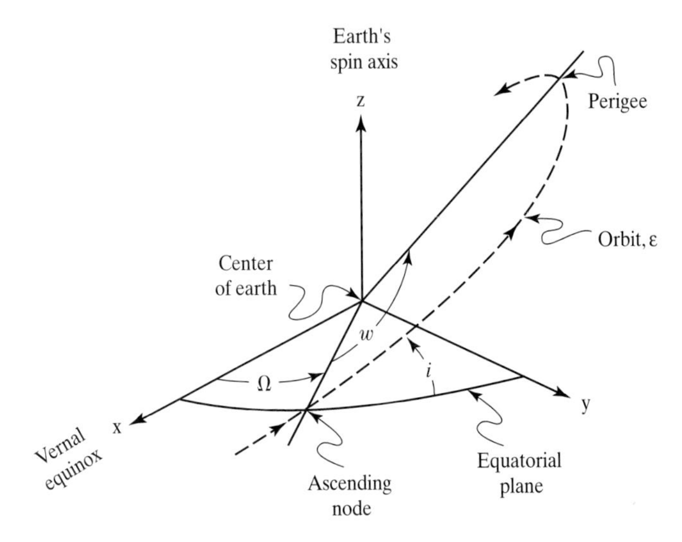


**Fig. 2.10** *A description of a satellite orbit. (Adapted from Kidder and Vonder Haar, 1995.)*

a, and is inclined by an angle, i, with respect to the equatorial plane of the earth. The point where the satellite passes the equatorial plane going north is the ascending node, $\Omega$, and the angle between the ascending node (equatorial plane) and the perigee (i.e., the point where the satellite's orbit takes the satellite closest to the earth) is known as the argument of perigee, $w$. The position of a satellite on its orbit (usually elliptical) at a particular time, assuming constant motion throughout the orbit is termed the mean anomaly, M. Simply,

$$M = n(t - t_p), \tag{2.23}$$

where $t_p$ is the time of the perigeal passage, and $M = 0$ at the perigee; n is the mean motion constant and equals $2\pi T^{-1}$, where T is the period of the satellite orbit. The orbital elements, $\varepsilon$, a, i, $\Omega$, $w$, M, and time are collectively known as the classical orbital elements and are necessary to determine the orbit of the satellite. If the classical orbital elements are constant, except for M, then the satellite's orbit is called Keplerian, or elliptical with the center of the earth as a focus.

Most meteorological satellite orbits are either *sun-synchronous* or *geosynchronous.* A sun-synchronous orbit has an a = 7,228 km, $\varepsilon = 0$, and i = 98.8°, and the sun-earth ascending node angle is constant. A satellite put into a sun-synchronous orbit passes the same spot at each latitude at the same time each day. The orbital plane of a sun-synchronous orbit rotates (precesses) approximately 1 degree eastward each day in order to keep pace with the earth's revolution around the sun. A geosynchronous orbit is synchronous with the rotation of the earth, has an i = 0, and an $\varepsilon = 0$. Thus a satellite in a geosynchronous orbit sits over the same position of Earth providing views, or scenes, of the Earth and its atmosphere below. The most popular example of the latter are the infrared and visible images of the weather shown on TV weather segments, or in the newspaper, and now over the Internet.

The radiometer on a satellite sees a scene that is bounded by a surface arc which is proportional to the distance, d, between the scan spot and the subsatellite point (Fig. 2.11). The scene or field of view of the satellite is twice d, and is called the swath width, $d_{sw}$, of the radiometer, and

$$d_{sw} = 2d = 2\phi r_e. \tag{2.24}$$

The radiation received at the satellite's radiometer(s) is expressed in terms of radiant energy, and may be represented as radiant flux density, dQ/dt, with units of Joules/s (See Equation 2.1). The radiant flux density that emerges from an area is also termed radiant excitance, and the radiant flux density incident on an area is commonly termed irradiance. The directional dependence of radiation is expressed as a function of the angle that portends the area of a projected object on the surface of a perfect square (Fig. 2.1b). This angle is termed the solid angle and is commonly denoted $\omega$ as previously defined. For example, $\omega = 2\pi$ steradians for an infinite plane, or $= A_c r^{-1}$ for an object with cross-sectional area, $A_c$, at a distance r from a point, and $A_c \ll r$.

Given that an object emits a radiant flux density per unit solid angle, or radiance, L, and the satellite detects a radiant excitance, E, that is given by

$$E = \int_0^{2\pi} L(\Theta, \phi) \cos\Theta \, d\omega, \tag{2.25a}$$

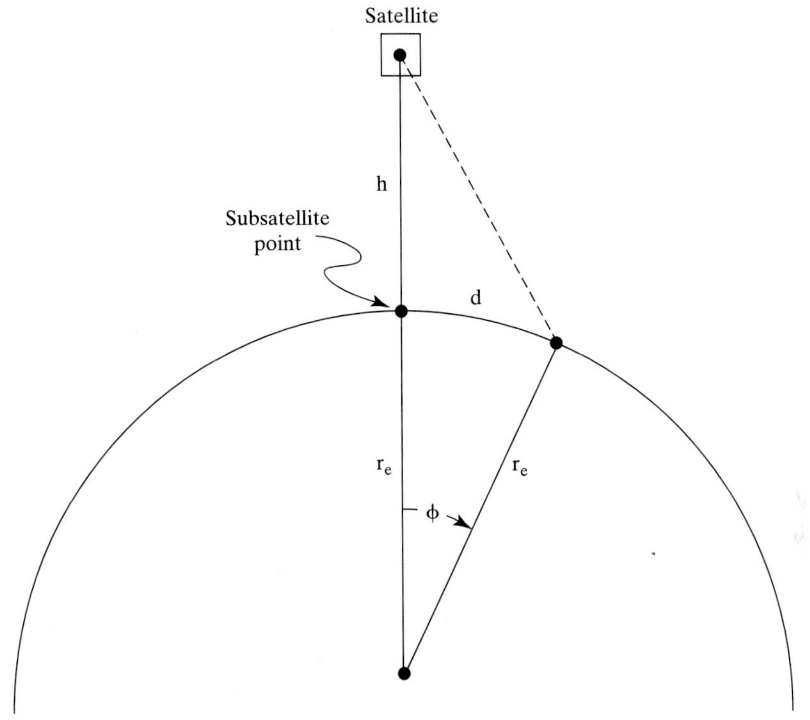


**Fig. 2.11** *The geometry of a satellite instrument earth scan. $r_e$ is the radius of the earth. (Adapted from Kidder and Vonder Haar, 1995.)*

where $L(\Theta, \phi)$ represents the component of radiation coming from an area perpendicular to the beam, and the $\cos\Theta$ takes into account all directions. Substituting in for $d\omega$

$$E = \int_0^{2\pi}\int_0^{\pi/2} L(\Theta\ \phi)\cos\Theta\sin\Theta\, d\Theta\, d\phi \tag{2.25b}$$

If the radiation is uniform in all $d\omega$ (i.e., it's isotropic radiation), then $E = \pi L$, similar to Equation 2.3, but now expressed with respect to the satellite. The radiance is also a function of wavelength (frequency via $\nu = c\lambda^{-1}$) and may be expressed in a similar form to Equation 2.4.

Radiation can either be absorbed, scattered or reflected as it passes through the atmosphere. The reflectance, $r_\lambda$, absorbance, $\alpha_\lambda$, and transmittance, $t_\lambda$, are the actual amount of reflected, absorbed, and transmitted radiation at wavelength, $\lambda$, divided by the total incident amount of radiation at $\lambda$, respectively. These three components when added together equal unity

$$r_\lambda + \alpha_\lambda + t_\lambda = 1. \tag{2.26}$$

Now let's briefly address the transfer of electromagnetic radiation through the atmosphere. This may be accomplished by considering radiation incident on a differential element of material, say our atmosphere with radiance, $L_\lambda$, as shown in Fig. 2.12. Radiance and its change are commonly used when discussing radiation

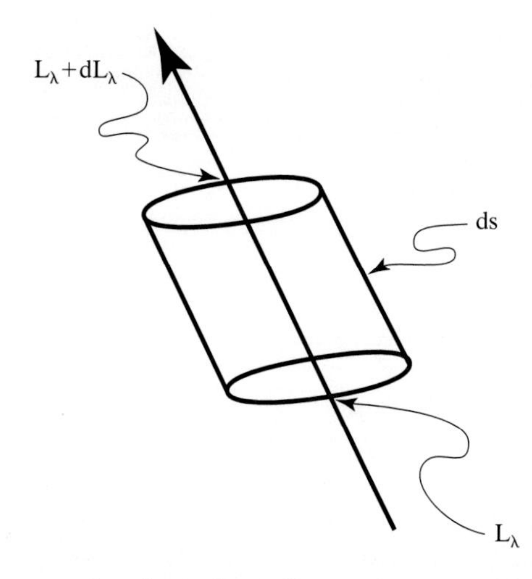


**Fig. 2.12** *Differential volume element of atmosphere wherein a beam of radiation that passes through it is altered. (Adapted from Kidder and Vonder Haar, 1995.)*

transfer since the radiance from an object would not change if there were no material in volume through which the radiation travels. Polarization effects on the transfer of the radiation through our atmosphere will be ignored. Thus, there are only four factors that affect the radiation beam as it travels through the atmosphere toward our satellite: The radiation emitted by the source material, arbitrarily denoted as K; the radiation scattered into the path of the radiation from other directions, arbitrarily denoted as K′; the direct radiation from the source that can be absorbed by objects in its path, arbitrarily denoted as E; the radiation scattered out of the radiation beam into other directions, arbitrarily denoted as E′. The change of radiance with distance, $dL_\lambda/ds$, is then

$$dL_\lambda/ds = (K + K') - (E + E'). \qquad (2.27a)$$

where, K and K′ are source terms since they add radiation to the beam, and E and E′ are depletion terms since they remove radiation from the beam. Expressions for K, K′, E, and E′ may be obtained by applying Kirchoff's law for the unprimed quantities and Beer's law for the primed quantities. Beer's law states, for example, that the intensity of the radiation diminishes as it travels from the top of the atmosphere toward the surface of the earth. The magnitude of the diminished intensity or extinction of the radiation as it passes through a medium is non-linearly proportional to the original intensity of the radiation, or

$$L_\lambda = L_{\lambda 0} e^{-\mu_\lambda} \qquad (2.27b)$$

where $L_{\lambda 0}$ is the initial intensity or radiance of the radiation with wavelength $\lambda$, $L_\lambda$ is the intensity of the radiation with wavelength $\lambda$ measured at a particular location "down stream" of the radiation source, $\mu_\lambda$ is the optical thickness of the layer, and the fraction of the diminished radiation is simply given by $e^{-\mu\lambda}$. Equation 2.27a may be written in a number of different forms depending on the application, all of which can be found in radiative transfer and recent satellite meteorology texts, such as Kidder and Vonder Haar (1995).

The earth's surface, the atmosphere, and the constituents within the Atmosphere (i.e. aerosols, clouds, and precipitation) all radiate. Radiation with a temperature-dependent intensity is detected by the satellite radiometer(s) after traveling through the atmosphere, and is generally converted into the temperature of a blackbody with similar radiant intensity, termed the brightness temperature. The brightness temperature, $T_b$, is defined in terms of the radiance and antenna temperature as

$$T_b = L_\lambda(T_a)(c_2(c_1)^{-1})\lambda^4, \tag{2.28a}$$

where $L_\lambda(T_a)$ is the blackbody radiance of an object radiating at temperature $T_a$, $\lambda$ is the wavelength, $c_1 = 1.191 \times 10^{-16}$ W m$^2$ sr$^{-1}$, and $c_2 = 1.4388 \times 10^{-2}$ m K. If the radiance or brightness of an emitter is linearly proportional to its temperature (i.e., Rayleigh–Jeans approximation) over the entire path viewed by the radiometer through an isothermal atmosphere, then $T_b$ may be expressed analytically in terms of the volumetric absorption coefficient of the radiators in the viewed path, $\sigma_a(\lambda)$, and the transmittance of the radiation through the viewed path, $\tau_\lambda$, as

$$T_b = \int_a^b T_a \sigma_a(\lambda)\, \tau_\lambda\, ds, \tag{2.28b}$$

where $\lambda > 1$ cm, $T_a$ is the temperature of the radiometer's sensor, or the antenna temperature, $\tau_\lambda$ is a function of distance outwards from the radiometer (point a) toward the surface (point b) $= \exp[-\tau(a,b)]$, and the optical depth, $\tau(a,b) = \int_a^b \sigma_a(\lambda)\, ds'$. $\sigma_a(\lambda)$ is the wavelength-dependent absorption cross-section of the radiator, which may be related to the target fall rate and may theoretically be expressed in terms of target size for a given $\lambda$ as long as the functional relations between target fall speed and target number density are known. For example, if the target is rain, then the target fall rate is the rain rate, which may be expressed in terms of raindrop diameter as long as the functional relations between the raindrop fall-speed and number density are known (e.g., Equation 2.18).The rain rate may not be known and it might be more practical to use the liquid water content (e.g., see Equation 2.21). However, even the liquid water content is related to hydrometeor size.

When the surface is taken into account, for example in the case of an isothermal precipitation layer (Kidder and Vonder Haar, 1995), then

$$T_b = T_p\left\{1 + \left[\varepsilon\left(\left(\frac{T_s}{T_p}\right) - 1\right)\tau_\lambda\right] - [(1-\varepsilon)\,\tau_\lambda^2]\right\} \tag{2.29}$$

where $T_p$ = antenna temperature of the precipitation layer that is assumed isothermal, $\varepsilon$ = emissivity of the surface and is assumed constant, and $T_s$ = temperature of the surface (°K), and the radiative contribution from space has been ignored. Note, that the assumption of an isothermal precipitation layer may not be valid, and the Rayleigh–Jeans approximation holds well in the microwave region of the electromagnetic spectrum. The brightness temperature is also used in the infrared region of the electromagnetic spectrum, but it is termed the equivalent blackbody temperature.

The brightness temperature derived from satellite data is generally treated as a simple linear transformation of $T_a$ into $T_b$, namely

$$T_b = (A_{vp} T_{av}) + (A_{hp} T_{ah}) + A_{op}, \tag{2.30}$$

where $T_{av}$ and $T_{ah}$ are the antenna temperatures for the vertical and horizontal polarizations of the electromagnetic radiation of a specified wavelength, respectively, and the A's are coefficients of the transformation (Wentz, 1988).

In summary, the fundamentals for measuring radiation from stationary and mobile platforms using traditional and non-traditional devices have been presented. These devices include a pyrheliometer, pyranometer, radiometer, radar, and lidar. We will draw upon the fundamentals in this chapter as we discuss the remote sensing devises designed to obtain measurements of temperature, windfield, water vapor, clouds, and precipitation, for example.

## *Exercises*

1. Derive the Wein displacement and Stefan-Boltzmann laws.
2. Calculate the maximum wavelength of a blackbody having a temperature of 200 °K, 290 °K and 3,000 °K.
3. Calculate the blackbody radiance for each of the wavelengths in Exercise 2 at a temperature of 290 °K. Repeat for 200 °K, and 3,000 °K.
4. What is the change in the amount of radiation emitted by the earth's surface if the earth's average surface temperature (assume 288 °K) were to increase to 290 °K? Repeat your calculation for an increase in the earth's surface temperature to 330 °K? Assume blackbody conditions in both cases.
5. Identify the primary absorbing species for the following absorption bands or wavelengths. Then identify the use or application for these bands or wavelengths.

| $\lambda$ ($\mu$m) | primary absorbing species |
|---|---|
| 0.69 | |
| 11.11, 3.76 | |
| 15, 4.3 | |
| 6.7, 6.3 | |
| 9.6 | |

6. Aliens from planet IR are very sensitive to infrared radiation, and in 2059 they happen to land on our planet and discover that they have a taste for eating humanoids. The normal temperature of humanoids is 29.6 °C. What wavelength of infrared radiation should these aliens be tuned to? What problems are they likely to incur at this wavelength? Assume that only the normal temperature of humans has changed, while the atmosphere has not changed from the present day atmosphere in any way. (Note: please ignore all problems except those which may be due to the atmosphere and its constituents).

CHAPTER THREE

# Temperature Measurement

The measurement of air temperature dates back to the time of Galileo (1564–1642) and is perhaps the premier measurement of meteorological parameters. Nonetheless, temperature measurement is not free from problems and one must be especially concerned with response time, accuracy, and operational need.

For example, if a fast response (i.e., time constant on the order of tenths of seconds to seconds) is required, then one might use an electronic sensor. A highly accurate and fast response sensor would be required on fast moving platforms. Actually there is an additional measurement concern that occurs in the vicinity of a sensor operating on fast moving platforms known as dynamical heating. On the other hand, if a slower response is acceptable (i.e., time constant on the order of a couple of minutes) then a mercury-in-glass thermometer might do. An instrument with a slow response could act as an integrator if the operational need is related to a Eulerian platform system.

Standard temperature measuring devices can provide 0.2 to 1 °C accuracy, while research grade devices can provide ≈0.01 to 0.2 °C accuracy. Modelling efforts should use data that are as accurate as possible, and in fact at least ten times more accurate than the best research devices. An error in temperature by as little as 0.2 °C can change the prediction of a global climate model. A ±1 °C error in temperature will result in a pressure error of ±0.01 mb. Our everyday need for temperature accuracy is ≈±2 or 3 °C so that we can get an idea of what clothing we might need to wear, and in fact the mercury or alcohol-filled thermometer works well for this purpose.

There are three kinds of temperature sensors, namely, mechanical, electronic, and passive or remote. Mechanical thermometers, which include mercury or alcohol-filled thermometers and bimetallic thermometers, can be good working standards and are widely used by industry, the weather enthusiast, and for some educational purposes. The electronic and remote sensors are superior to the mechanical sensors in a number of ways, but keep the following three things in mind: (1) mechanical thermometers may have better stability than some electronic devices; (2) some of the mechanical thermometers are capable of yielding near research temperature accuracies; and

(3) the operational needs of the measurement campaign. The following presents the fundamentals of the electronic and passive or remote temperature sensing devices. Middleton and Spilhaus (1953), BMOAM (1961), and Brock (1984) contain additional details on liquid-in-glass thermometers.

## 3.1. Electronic Thermometers

Thermocouples, resistance elements, and thermistors are the three basic kinds of electronic thermometers. They operate under the principle of measuring changes in electric signals as a function of temperature, and are generally discussed with respect to the two ways for which they yield their temperature measurement, namely: (1) self-generating (e.g., thermocouples), and (2) non-self-generating (e.g., resistance elements, thermistors, and diodes). The self-generating electronic temperature devices produce an electric current as a function of temperature. The non-self-generating require the application of an external signal in order to detect a change in a property such as resistance.

***3.1.1. Thermocouples.*** Thermocouples (Fig. 3.1) are accurate, highly reliable, rugged, easily-installed, low-cost devices that work over a wide temperature range, and are compatible with most measuring and recording systems. They consist of two different metals joined together to make a continuous circuit and operate according to a principle known as the *thermoelectric effect*. Simply put, when one junction has a different temperature than the other, an electromotive force is produced in the circuit and current flows. The magnitude of the force or potential depends on the temperature difference between the two junctions. If one of these junctions is kept at a constant temperature (i.e., the reference temperature) and the other is allowed to obtain the temperature to be measured, then the electric potential created in the circuit gives a measure of the temperature difference between the two junctions (one is usually colder than the other), and the required temperature may then be determined. The electric potential is measured by connecting a galvanometer into the circuit, or a potentiometer if more accurate work is required. There are three components of the thermoelectric effect, namely the Seebeck effect, Peltier effect and the Thompson effect.

The Seebeck effect is the conversion of thermal energy to electrical energy. This effect "measures" the ease at which excess electrons will circulate in an electrical circuit under the influence of thermal difference. The change in the voltage (through a circuit of two dissimilar materials) is proportional to the temperature difference between the junctions when the ends are connected to form a loop,

$$\partial E = N_{A,B} \partial T \tag{3.1}$$

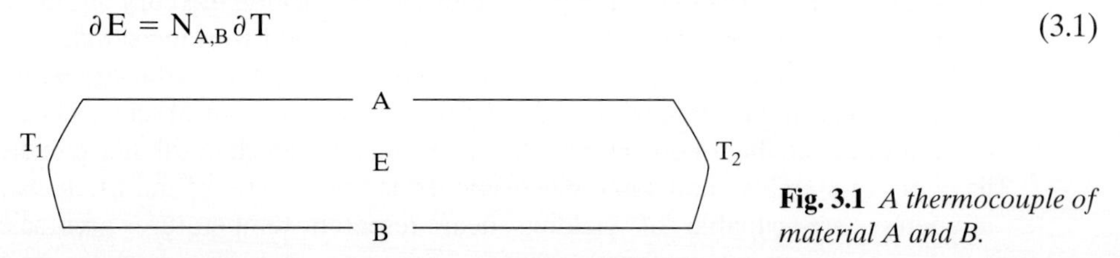


**Fig. 3.1** *A thermocouple of material A and B.*

**Table 3.1** *Seebeck coefficients for various thermocouples*

| *Thermocouple Type* | *Seebeck Coefficient* [$\mu V/°C$][a] |
|---|---|
| J | 50.2 {51 (5.1)} |
| K | 39.4 {40 (4.0)} |
| N | 26.2 {not available} |
| E | 58.5 {62 (6.2)} |
| T | 38.0 {40 (4.0)} |

[a]Bracket Values were obtained at 20 °C at a digital voltmeter sensitivity for 0.1 °C; Otherwise values are at 0 °C. The number in parentheses is interpreted as the voltmeter resolution in ($\mu V$) to detect a 0.1 °C change at 20 °C; for example, 4 $\mu V$ resolution if a Type T thermocouple is used to detect a 0.1 °C change.

where E is the thermoelectric force, and $N_{A,B}$ is the Seebeck coefficient (see Table 3.1), or the thermoelectric power of the metallic combination A and B. The thermoelectric force, EMF, may be approximately expressed as a non-linear function of temperature with the general form,

$$E = \alpha T + \beta T^2 + \ldots + \eta T^n \qquad (3.2)$$

where the coefficients $\alpha$ and $\beta$ are constant for a given material (with respect to platinum), T is the temperature of the measuring junction in °C, and the temperature of the reference junction is 0 °C. Table 3.2 provides magnitudes of the coefficients $\alpha$ and $\beta$ in Equation 3.2 for some metals. Platinum is the standard for measuring the Seebeck coefficient.

The Peltier effect is closely related to the Seebeck effect. It represents the thermal effect due to a reversible current through dissimilar materials (regardless of whether the current was from an external source or an internal source), or through similar metals due to an external source of current (Fig. 3.2). That is, a current flow in one direction might warm the junction of the two dissimilar materials (and release

**Table 3.2** *$\alpha$ and $\beta$ coefficients (with respect to platinum) for determining thermocouple EMF given by Equation 3.2. (Adapted from Wang, 1975; OMEGA, 1992)*

| *Metal* | $\alpha$ ($\mu v$ °C$^{-1}$) | $\beta$ ($\mu v$ °C$^{-2}$) |
|---|---|---|
| Iron (Fe) | +16.7 | −0.0148 |
| Copper (Cu) | + 2.7 | +0.0040 |
| Constantan (Cu + Ni) | −34.6 | −0.0279 |

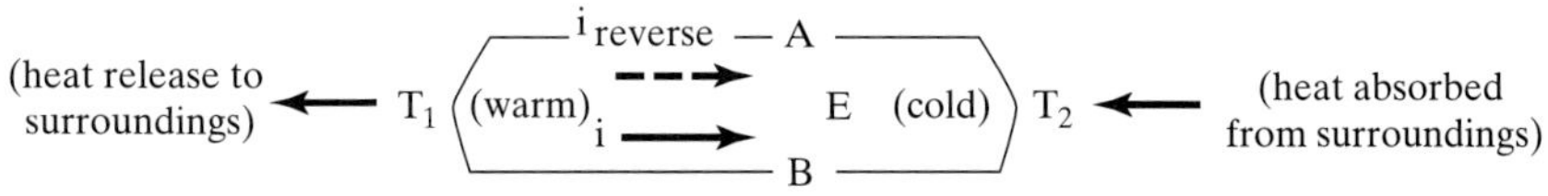

**Fig. 3.2** *The Peltier effect on material A and B.*

heat to the surroundings of that junction), whereas, if the current was reversed, the junction would cool (and absorb heat from its surroundings).

The Thompson effect is the absorption or liberation of heat by a homogeneous conductor (that has a temperature gradient within it) due to a current flowing through it. It is primarily evident in currents introduced from external sources and those generated by the thermocouple itself. The ability of a given material to generate heat with respect to both a unit temperature gradient and a unit current, is gauged by the Thompson coefficient.

The importance of the Peltier and Thompson effects is essentially infinitesimal because the heat evolved is negligible compared to the amount of thermal energy available from the environment to the junctions at $T_1$ and $T_2$. However, precision measurements use high impedance circuits that virtually eliminate the Peltier and Thompson effects.

**Thermocouple (Thermoelectric) Laws** The three fundamental empirical laws behind the accurate measurement of temperature by thermoelectric means are the (I) law of homogeneous materials, (II) law of intermediate materials, and (III) law of intermediate temperatures. They are highlighted as follows:

1. The application of heat will not sustain an electric current in any circuit of a single homogeneous material of any size.
2. The E of a two material circuit remains unaffected by the addition of a third material, C, as long as the temperature of the circuit remains constant. Furthermore, given a circuit with a constant temperature, the algebraic sum of the thermally induced electromotive forces in multi-material (i.e., more than one metal) circuits is zero.
3. The electromotive forces of any number of individual thermocouples joined at similar temperature junctions are additive. However, the reference temperature of the set must be known in order to determine the magnitudes of E. The coefficients, $\alpha$ and $\beta$, (Table 3.2) for each individual non-platinum metal used may be algebraically combined (e.g., Wang, 1975).

---

**Example 3.1**

Suppose, a three thermocouple set is used to measure the air temperature. One of them has two dissimilar homogeneous metals that produce a thermal emf of $E_{12}$ when the junctions are at temperatures $T_1$ and $T_2$, and another has a thermal emf of $E_{23}$ when the junctions are at temperatures $T_2$ and $T_3$. What is the thermal emf of the third thermocouple, $E_{13}$?

The thermal emf, $E_{13}$, generated when the junctions are at temperatures $T_1$ and $T_3$ will be $E_{12} + E_{23}$.

---

**Example 3.2**

Assume that an iron-copper thermocouple registers temperatures of 0 °C and 200 °C at its reference junction and the measuring junction, respectively. What are the values of the combined $\alpha$, $\beta$ coefficients? What is E for this thermocouple? (Assume 2nd order in T).

$$\alpha_{Fe-Cu} = \alpha_{Fe} - \alpha_{Cu} = 16.7 - 2.7 = 14 \mu v\ ^\circ C^{-1}$$

$$\beta_{Fe-Cu} = \beta_{Fe} - \beta_{Cu} = -0.0148 - 0.0040 = -0.0188\ \mu v\ ^\circ C^{-1}.$$

Recall,

$$E = \alpha T + \beta T^2 = 14\ \mu v\ ^\circ C^{-1}(200^\circ C) + (-0.0188\ \mu v^\circ C^{-1})(200^\circ C)^2$$
$$E \approx 2048\ \mu v.$$

Note: (1) The National Bureau of Standards (Powell et al., 1975) has listed the aforementioned coefficients and electromotive forces for individual thermocouple types rather than individual metals. The reference temperature used is 0 °C. (2) The first metal given is the positive wire, and the other is negative. For example, consider an iron-copper thermocouple. The positive wire is iron and the negative wire is copper.

---

Thermocouples may be configured in series or in parallel depending on the desired application (Fig. 3.3). Two or more (typically 3) thermocouples connected in series, with all of the measuring junctions at $T_2$ and their reference junctions at $T_1$, make up the most common thermopile. The potential, E, of a thermopile composed on n thermocouples of potential $E_{ab}$, where a,b denote the material of the thermocouple per above, is given by

$$E = \sum_{i=1}^{n} (E_{ab,i}).$$

The electromotive force, E, generated from n thermocouples connected in parallel is the same as that for the individual thermocouples. Consequently, if all of the thermocouples are of equal resistance, then E will correspond to the average of the temperatures at the individual junctions.

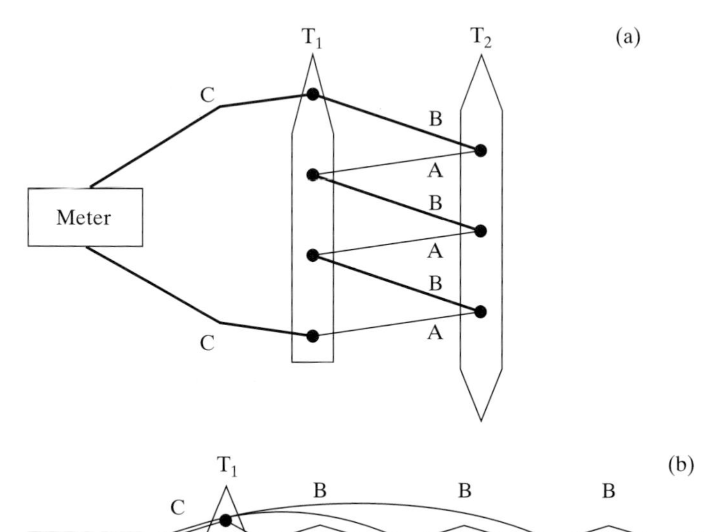


**Fig. 3.3** *Some common configurations of thermocouples; (a) series, (b) parallel.*

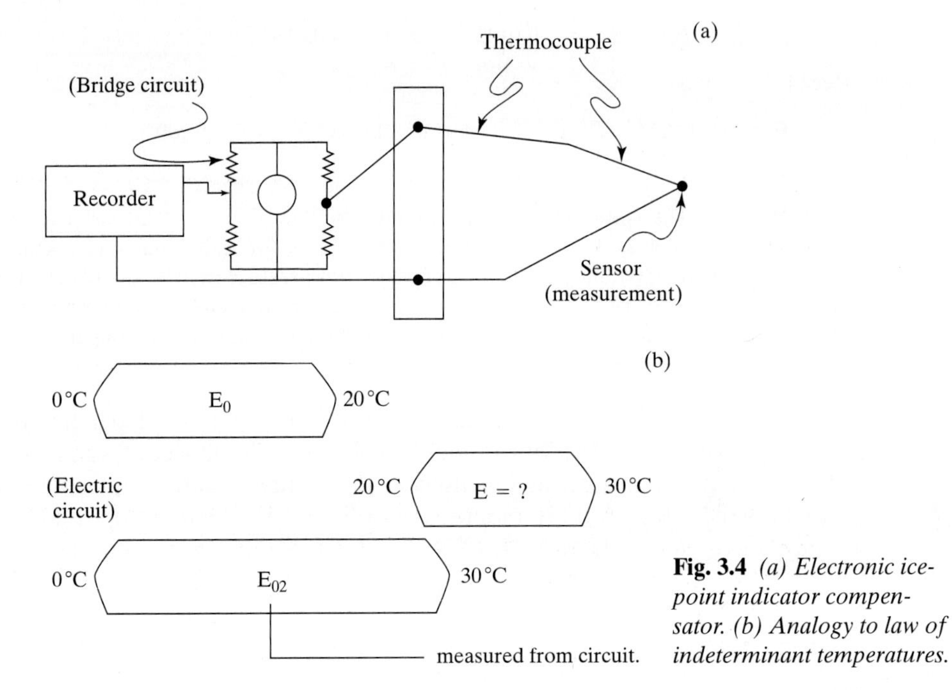


**Fig. 3.4** *(a) Electronic ice-point indicator compensator. (b) Analogy to law of indeterminant temperatures.*

The thermocouple reference junction must be maintained at a constant temperature when accurate thermocouple measurements are necessary, otherwise any readings made will be in error. The value of the constant reference temperature is usually the melting point of ice (0 °C). An electronic ice-point indicator compensator (Fig. 3.4a) is typically used to provide the reference temperature, and contains a bridge circuit that generates a signal of opposite polarity and equal magnitude to the thermocouple reference junction attached to the active leg of the bridge. The latter improves the error of the copper-constantan thermocouple to ±0.2 °C between the temperature range of 10 °and 40 °C. Of course, an ice bath might also be used for the reference temperature.

The operation of the electronic ice-point compensator is analogous to the law of intermediate temperatures (Fig. 3.4b). For instance, consider a thermocouple with an electromotive force, $E_0$, and temperatures of 0 °C at its reference, and 20 °C at its measuring junction. The measuring junction of the thermocouple is simply connected to an electric circuit with the same temperature. This circuit makes a measurement of 30 °C. The E of this individual circuit is unknown, but the electromotive force of the entire set of devices is what is measured, namely $E_{02}$. In micrometeorological practice, where the variance, $\sigma_T^2$, or the covariance of temperature with the vertical velocity (i.e., $< w'T' > = \sigma_{wT}^2$) is a desired measurement, the reference junction is simply imbedded in a block of metal whose temperature does not vary in a matter of tens of minutes.

The general operating temperature range of thermocouples is quite large (−273 °K to 6,000 °K) and is size and type dependent. Note the size of the measurable temperature domain compared to that of the conventional thermometer. The choice of which thermocouple type to use depends on desired error limits, response time, and measurable temperature domain among others.

There are many different combinations of metals that are used to make thermocouples. The metals that are commonly used for temperature measurements include silver, copper, platinum, nickel, and iron. Common alloys include constantan (60% copper and 40% nickel), manganin (a mixture of copper, manganese, and nickel), and nichrome (nickel-base alloy with chromium and iron). Platinum is the most accurate resistance thermometer sensor. It is highly stable, durable and pure. Furthermore, it resists contamination and corrosion and consequently does not drift with age and use. It is also very expensive and other metals and alloys are more widely used. Some of the more commonly used thermocouples are given in Table 3.3 along with their symbols (as used in the reference tables), standard and special limits of error, temperature range, and primary application. A more complete Table may be found in Powell et al. (1975), or OMEGA (1992).

A commonly used thermocouple for meteorological applications is the copper-constantan (T-Type). Its range of use varies from 350 °C down to −200 °C, but it is most frequently used within the temperature range from −60 °C to 100 °C with an accuracy of ±0.5 °C or 0.4% (whichever is greater). The T type thermocouple is good for moist environments.

Another common thermocouple is the iron-constantan (J-Type), although it is not wisely used for meteorological purposes since its output ranges from 0 °C to 750 °C and iron tends to rust. It is most frequently used to measure the 0 °C to 200 °C temperature range with an accuracy of ±1.1 °C or 0.4% (whichever is greater). Its low cost, and the high melting point of iron compared with that for copper make it extensively used in industry. The infrared thermocouple is relatively new, and simply generates its signal by converting the radiated heat energy it receives to an electrical signal that is amplified to a standard thermocouple Type configuration (OMEGA, 1994, 1995). The ranges specified above may be varied to improve the measuring accuracy compared to the values indicated in Table 3.3. Regardless, the stated accuracies may be improved by enhancing the design of the electronic circuitry in which the thermocouple is placed.

***3.1.2. Resistance Thermometers.*** There are two kinds of resistance thermometers, namely positive and negative. Positive resistance thermometers, or positive resistance elements, are made of pure metals and alloys whose resistance increases with increasing ambient temperature. Negative resistance elements are made of metals and alloys that show an inverse relationship between resistance and temperature. For example, conductors (e.g., copper, Cu; platinum, Pl) are typically used to create positive resistance elements and have positive resistance coefficients while semiconductors (e.g., thermistors, transistors, diodes) are typically used to create negative resistance elements and have negative resistance coefficients. A typical resistance coefficient is ≈0.4 ohm (Ω) $°C^{-1}$, especially for a 100 Ω platinum element.

Resistance thermometers are based on the principle that the electrical resistance of a material varies with temperature. They are very fragile and, consequently, are commonly used as parts of other measuring devices, such as a wheatstone bridge

**Table 3.3** *The range and limits of error of some thermocouples. (Adapted from OMEGA, 1992, 4, 5)*

| *Thermocouple type* | *Temperature Range*[a] [°C] | *Error Interval*[b] *Standard* [°C (%)] | *Error Interval*[b] *Special* [°C (%)] | *Application* |
|---|---|---|---|---|
| J (Iron vs Constantan) | 0 to 200 | ±2.2 (±0.75) | ±1.1 (±0.4) | Recommended for vacuum or reducing applications. Industry, due to low cost, and relatively high Seebeck coefficient. |
| | 200 to 750 | ±2.2 (±0.75) | — | |
| K (Chromel vs Alumel) | −200 to 0 | ±2.2 (±2.00) | — | Multi-purpose, except not for vacuum or reducing applications |
| | 0 to 200 | ±2.2 (±0.75) | ±1.1 (±0.4) | |
| | 200 to 1250 | ±2.2 (±0.75) | — | |
| N (Nicrosil vs Nisil) | −270 to 0 | ±2.2 (±2.00) | — | Alternative to Type K; more stable at high temperatures. |
| | 0 to 200 | ±2.2 (±0.75) | ±1.1 (±0.4) | |
| | 200 to 1300 | ±2.2 (±0.75) | — | |
| E (Chromel vs Constantan) | −200 to − 0 | ±1.7 (±1.00) | — | Oxidizing environ. Highest EMF change per degree. |
| | 0 to 200 | ±1.7 (±0.50) | ±1.0 (±0.4) | |
| | 200 to 900 | ±1.7 (±0.50) | — | |
| T (Copper vs Constantan) | −200 to − 60 | ±1.0 (±1.50) | — | Good where moisture and low temperatures are present. |
| | −60 to 0 | ±1.0 (±1.50) | ±0.5 (±0.4) | |
| | 0 to 100 | ±1.0 (±0.75) | ±0.5 (±0.4) | |
| | 100 to 350 | ±1.0 (±0.75) | — | |
| Infrared[c] | −45 to 290 | ⩾±1.0 (±2.0)[d] | — | Ground temperature, maybe air temperature. |

[a] The actual temperature range might extend beyond that given, but the reliability of those measurements is very poor. The error also depends on wire thickness.

[b] These are maximum error intervals for two different thermocouple grades, standard and special, with special being better, over the entire range of temperatures listed to the left in this Table. Use the error interval that is greatest (i.e., deg. or %). Errors can be reduced for specific smaller temperature ranges, and a more accurate zero point or reference point, such as an electronic ice point indicator bridge circuits. For example, the error is ±0.2 °C for copper-constantan thermocouples for temperatures between 10 and 40 °C.

[c] Applicable for J, K, T, E thermocouple calibrations (OMEGA, 1995).

[d] The accuracy can be greatly improved with a microcomputer controlling the particular measuring device.

circuit. A small electric current is passed through the wire, and its resistance is commonly measured by connecting it to a wheatstone bridge circuit. The null-balance bridge circuit is preferred so that one can adjust the resistance of one of its resistors until zero current flows through the galvanometer. The unknown resistance may then be determined since all of the other resistances of the bridge are known. However, since changes in the resistance of the element will produce changes in the out-of-balance current measured by the galvanometer, it may be prudent to have the galvanometer calibrated directly in terms of temperature prior to its use in this case. The use of large electric currents will produce self-heating errors.

Resistance elements are calibrated by determining their resistances at 0 °C, 100 °C, and −78.51 °C (sublimation point of $CO_2$). The response of both conductors and semi-conductors to inputs is non-linear with respect to temperature, T. The general form of the response, $R_T$, curve for conductors is given as

$$R_T = R_{T=0°C}\,[1 + f(T - 0°C) + \beta(T - 0°C)^2], \tag{3.3}$$

where $f$ and $\beta$ are the first and second order temperature coefficients, respectively.

The typical resistance thermometers are made with a 100 $\Omega$ resistance and very low magnitude outputs. The low magnitude outputs are a problem for a number of reasons, including the determination of whether the measurement is simply noise or a real measurement. Most noise problems are due to the application and design of the measurement systems and not only due to resistance thermometer characteristics. Nonetheless, the self-heating error is <0.1 °C for the typical resistance thermometer as long as the current within the element is 2 mA (Fritschen and Gay, 1979). A 2 mA current could be accomplished through good practice and design of the measurement system.

***3.1.3. Thermistors.*** The thermistor, or thermal resistor, is a hard, ceramic-like, electronic semi-conductor, commonly made from a mixture of metallic oxide materials. It has a very large negative resistance coefficient (i.e., an increase in T by 1 °C yields a 5% decrease in resistance), especially compared to resistance thermometers. Thermistors come in many sizes and shapes (e.g., bead, disc, rod).

There are three basic electrical characteristics of thermistors that are particularly noteworthy, namely, resistance-temperature, voltage-current, and current-time. The resistance of a thermistor is solely a function of its absolute temperature (°K). There is a necessity to use very small amounts of power when measuring the resistance of a thermistor since the thermistor can be heated above the temperature of its surroundings by the power dissipation, consequently lowering the resistance it indicates. The relationship between electrical resistance and temperature, T(°K) is given as

$$R(T) = R(T_0 = 298.16°K)\exp\{B(T^{-1} - T_0^{-1})\} \tag{3.4}$$

where R is the resistance at temperature T (°K), $R(T_0 = 298.16\ °K)$ is the resistance at $T_0$, and B is a constant depending upon material, also termed the "material constant", and is ≈4000 °K for some common thermistor materials (Wang, 1975). The definition of the temperature coefficient of a thermistor, $\alpha$, is derived from Equation 3.4 and is

$$\alpha = [dR(T_0 = 298.16°K)dT^{-1}]R(T_0 = 298.16°K)^{-1} \approx -BT^{-2} \tag{3.5}$$

The value of $\alpha$ may be expressed as ohms (ohms °K)$^{-1}$ or % °K$^{-1}$.

The voltage-current characteristic is simply the increase in current due to a gradual increase in voltage. The increased current will generate heat within the thermistor that could eventually increase its temperature above that of its surroundings, causing a reduction in resistance.

The current-time characteristic is simply the time period for current flow and is quantified by the time it takes a thermistor to reach its maximum possible temperature for a given amount of power available in the circuit. A quasi-steady state condition exists at the point of maximum temperature. Keep in mind that since the thermistor has mass, it will take a certain amount of time to reach maximum heating. Thus, the response time of a thermistor is a function of the mass of the thermistor, the resistance of the circuit, and the amount of voltage applied. This response time may be of the order of 0.001 seconds to several hours. In contrast, the IR thermocouple has a response time that is generally well within 0.3 s.

Thermistors are very similar to resistance thermometers in terms of accuracy. However, each individual thermistor has to be calibrated, and self-heating errors in thermistors can be quite large because of the large resistance coefficients (e.g., Fritschen and Gay, 1979). Thermistors are commonly used in circuits designed for temperature measurements involving telemetry, high sensitivity, and/or horizontal gradients (Fig. 3.5a). For example, they are used in hot wire anemometers to measure windspeeds.

Diodes, especially silicon of medium size, typically have a linear temperature coefficient of $\approx -2$ mV/°C. The primary problem associated with diodes is the difficulty in finding two diodes with the exact same features, which limits their use. Nonetheless, diodes are also highly conductive, have a low likelihood of suffering from self-heating, are relatively inexpensive, and are used in analog thermometric devices.

The variation of temperature with respect to voltage or resistance input is conceptualized in Figure 3.5b for a resistance element device, thermocouple, and thermistor.

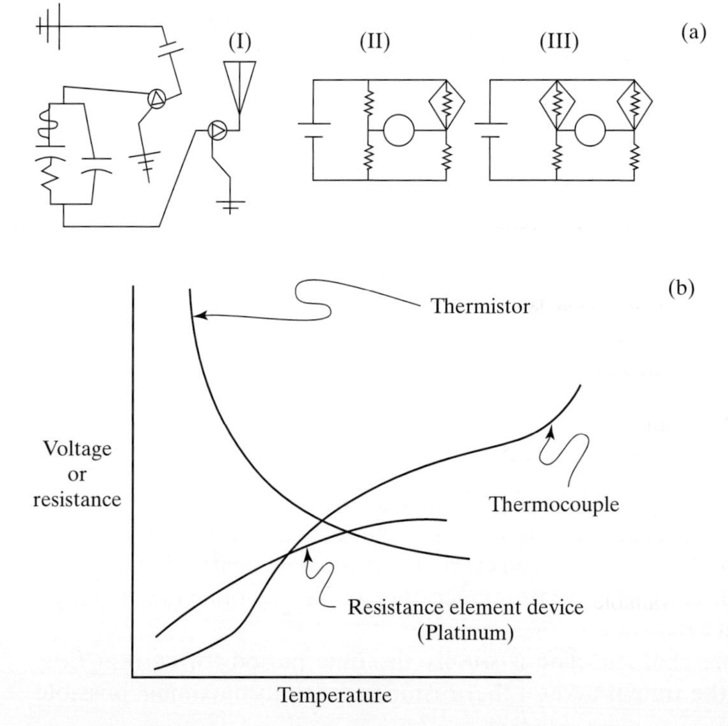


**Fig. 3.5.** *(a) Three common circuit configurations for measuring temperature with thermistors: (I) telemetering (transmission); (II) high sensitivity; (III) vertical or horizontal gradients. The diamond indicates the unknown or sensing resistor(s). The downward triangle is an antenna, and the circled triangle is an amplifier. (After Wang, 1975.) (b) Voltage or resistance versus temperature for a resistance element device, thermocouple, and thermistor.*

***3.1.4 General Comparison of Electronic Thermometers.*** Let us visit the question of what electronic temperature device should be used. Electric thermometers are chosen by at least considering: temperature range, accuracy, and sensitivity required for the application, and the following summarizes some useful notes in this regard. The advantages and disadvantages for each of the three types of electronic thermometers—including the recently introduced infrared thermocouple and the liquid-in-glass thermometer for comparison—are given in Table 3.4.

**Table 3.4** *Advantages and disadvantages of using thermocouple, IR thermocouple, resistance and thermistor temperature, T, measuring devices*

| *Device* | *Advantages* | *Disadvantages* |
|---|---|---|
| Thermocouple | Self-powered<br>Simple<br>Rugged<br>Many configurations<br>Wide temperature range<br>Small thermal mass, so thermal response<br>Inexpensive | Non-linear response (volt vs T)<br>Low voltage<br>Reference required<br>Least stable of the four[a]<br>Least sensitive of the four[a] |
| Infra Red (IR) thermocouple | No measurable long-term calibration change<br>Self-powered<br>Wide temperature range<br>Simple<br>Rugged<br>Can have standard thermocouple, resistance element, and thermistor outputs | Temperature represents that in a specific portion of the electromagnetic IR spectrum<br>Approximately same accuracy as standard thermocouple<br>Nearly 5 times as expensive as standard thermocouple<br>Highly non-linear response |
| Resistance element | Most stable of the four, except IR thermocouples<br>Most accurate of the four<br>Response (resistance, R vs T) is more linear than thermocouples | Expensive<br>Current source required<br>Low absolute resistance<br>Small $\Delta R$<br>Self-heating |
| Thermistor | High output<br>Fast response<br>Two-wire ohms measurement<br>Very accurate | Non-linear response (Resistance vs T)<br>Limited temperature range<br>Fragile<br>Current source required<br>Self-heating |
| Liquid-In-glass[b] | Readily available | Relatively long time constant (minutes) |

[a] The phrase "of the four" throughout this Table refers to the electronic devices identified in this Table.

[b] Added for comparison only.

Thermocouples have a generally wider temperature range than the positive resistance thermometers, the thermistors, and the typical liquid-in-glass thermometer (e.g., OMEGA, 1995). The thermistor is the most accurate ($\pm \leqslant 0.01$ °C) of the electronic thermometers, and also has the highest resolution. Platinum wires have an accuracy of $\approx \pm 0.1$ °C, and the special grade thermocouple has an $\approx \pm 0.5$ to $\pm 1.1$ °C accuracy. Thermocouples can be made with better accuracy. The accuracy at extremely high and low temperatures (i.e., at the edges of the individual sensor range) is limited for all electronic thermometers.

The sensitivity of electronic thermometers (especially resistance thermometers) may be gauged by either a percent change in resistance (i.e., % °C$^{-1}$), the output resistance per °F or °C over the selected temperature range (i.e., $\Omega$ °C$^{-1}$), or as relative sensitivity (i.e., $\mu$ volts °C$^{-1}$).

Thermocouples and thermistors are used as components of radiometers, dew point hygrometers, and psychrometers. Recall that thermistors are also used as components of anemometers. The infrared thermocouple may be used as a hygrometer, and might also find some application in measuring the actual surface temperature of the ground, and possibly even the air. Platinum wires and thermistors are used for direct measurements, compared to thermocouples who require a reference junction. Consequently, thermocouples require more complicated circuitry.

It is worth noting that a capacitive temperature sensor, called a THERMOCAP®, is used in the standard radiosonde units made by Vaisala (Vaisala, 1994). The THERMOCAP® (Vaisala, 1994) temperature sensor is a small capacitive bead encapsulated in glass, treated with a water repellant and a metallization treatment of the surfaces to ensure excellent performance in rain and minimum radiation sensitivity. The two temperature conductors sense the temperature and yield different resistances. The result is a change in the current flow that changes the capacitance of its circuit. The change in capacitance is telemetered to the surface-receiving station as a frequency. The latest Vaisala radiosonde temperature sensor consists of two 25 $\mu$m platinum wire capacitors, has a very quick time response (Fast F THERMOCAP®, Table 3.5), and is minimally effected by solar radiation (Ken Goss, Vaisala, personal communication, August 1996). Chapter 8 has some additional details of the Vaisala sensors.

## 3.2. *Remotely Sensed Temperature Measurement*

Remotely sensed temperature measurements do not require that the sensing device be in contact with the temperature source. The devices that are designed to make passive temperature measurements are infrared temperature measuring devices, radiation thermometers, and pyrometers. Pyrometers are devices that measure the intensity or spectral radiance of the intercepted thermal radiation, whereas radiation thermometers determine the temperature of an object by intercepting and measuring the thermal radiation it emits (OMEGA, 1994). The latest type of infrared temperature-measuring device is the infrared thermocouple that has been highlighted in the previous section. The pyrometers and radiation thermometers are very similar to radiometers. Consequently this subsection will be split into two—specifically discussing infrared temperature sensors and generally discussing

**Table 3.5** *Times constants of various thermometers*

| *Thermometer* | *Sensing element dimension* | *Ventilation m s^-1* | $\tau$ *s* |
|---|---|---|---|
| Mercury-in-glass | Sphere, 1.12 cm dia. | 4.6 | 56 |
| Mercury-in-glass as wetbulb | Sphere, 1.12 cm dia. | 4.6 | 52 |
| Mercury-in-glass | Sphere, 1.065 cm dia. | 4.6 | 50 |
| Electrical resistance, coiled element | Cylinder former, 3.8 cm dia. 3.5 cm long | 4.6 | 8 |
| Aspirated thermocouple | 4 thermocouple junctions | 11 | 1.8 |
| Aspirated thermocouple as wetbulb[a] | 4 thermocouple junctions | 11 | 2.7 |
| Capacitive temperature sensor (RS-80, Vaisala) | small bead | 6[b] | 2.5 |
| Capacitive temperature sensor (RS-90, Vaisala)[c] | small bead | 6[b] | 0.2 |
| Generic Infrared Sensor | | — | 0.2–5.0[d] |

[a] Good wetbulb will have a smaller $\tau$ than the dry bulb.

[b] Based on surface tests.

[c] Fast F-THERMOCAP® temperature sensor by Vaisala. RS denotes radiosonde, and RS-80 is Vaisala's standard radiosonde.

[d] Typical time constants are ≈0.5 s, although 0.01 s time constants may be obtained with silicon detectors.

temperature measurements made by radiometers. The fundamental principles behind these devices have been discussed in Chapter 2.

Infrared thermometers come in a multitude of configurations that complement optical, electronic, technology, size, and even protective enclosures (OMEGA, 1994, 1995). Nonetheless, they all convert infrared energy input into an electronic output that is then relatable to the temperature of the emitting surface. The processing of the input into output follows a similar plan to that of the radiometer (Figure 2.2), and the exact configuration depends on the particular application (e.g., OMEGA, 1992, 1994, 1995). For example, the IR detector has a field of view, or scene, that ranges from ≈0.5° to 90° and a minimum detectable spot size that ranges from >≈0.01 cm at a distance very near the detector and beyond. The detector may be made of silicon, Ge, or lead sulfide, PbS, especially for applications between 0.5–1.0 μm, 0.7–1.7 μm, or between 0.5–3.0 μm, respectively. The most common commercially available spectral region and emissivity range are between 4–20 μm and 0.8–1.0 respectively, although the latter may be chosen to suit. The output from the detector is generally non-linear, and of the order of 100–1,000 μV. The signal is amplified to the desired thermocouple (i.e., J, K, N, etc.) characteristics, regulated, and linearized. Ultimately, the analog output is a linear millivolt or milliamp signal, that is transferred to an RS-232, or sent to a remote display or recorder. The addition of a microcomputer to the aforementioned analog infrared temperature device increases the accuracy of the infrared temperature measurement, replaces the linearization techniques, provides automatic calibration as well as built in datalogging capabilities. For example, the OMEGA OS5100-C (OMEGA, 1995) has an operating temperature range of 0 to 65 °C in the spectral region of 8–12 μm, variable emissivity from 0.1 to 1.0, a response time of

84 milliseconds for 95% complete response, and a minimum detectable spot size of 0.08 cm at 0.8 cm away from the sensor for a few thousands of dollars, including complete setup and control software.

Some IR temperature-measuring systems use fiber optics to detect the input when conventional infrared detectors are not possible (e.g., OMEGA, 1994, 1995). Infrared detectors are primarily used for industrial and military applications, but they may also be used for humidity measurement, surface temperature measurement, and possibly air temperature measurement. While these sensors are reasonably accurate they are not yet ready to replace the mercury or alcohol thermometers for air temperature measurement.

The most popular platform used to make temperature, vapor, and windfield soundings today is the radiosonde. The radiosonde may ultimately be replaced by satellite-derived soundings of temperature, water vapor, and the windfield. Satellite-derived temperature soundings are fundamentally different from radiosonde soundings, and the agreement between the two is very good for clear air situations. The generic radiosonde soundings have a root mean square error of 1 °K. The satellite-derived temperature soundings do not have the same resolution as the radiosonde soundings, and they yield maximum deviations from the radiosonde sounding in regions where (i) the temperature lapse rates vary greatly with height (for example, near the tropopause and the surface), and (ii) the atmosphere is cloudy. Surface-based microwave radiometric measurements of the vertical temperature profile of the atmosphere are in better agreement with the profiles derived from radiosonde than those from a satellite platform (e.g., Snider, 1971; Westwater, 1972; Kidder and Vonder Haar, 1995). Please keep in mind that radiometers are still very expensive.

Infrared interferometers will soon be able to operationally provide greater vertical resolution in temperature, water vapor, and other trace gas concentration soundings. Smith et al. (1987) discuss a system, named the high-spectral-resolution interferometer sounder, that has 0.3 $cm^{-1}$ spectral resolution and is presently under development. Kidder and Vonder Haar (1995) suggest that significant improvement in the vertical resolution of temperature soundings might not come until the lidar or some other active sensor, becomes feasibly applied in space, which may soon happen. The retrieval of surface temperatures from interferometer data has been successful over sea (e.g., McClain et al., 1985) and land (e.g., Price, 1984) compared to satellite radiometer data. Smith et al. (1996) measured the air temperature just adjacent to the sea surface with an interferometer from an oceanographic ship with an accuracy believed to be better than 0.1 °C.

## 3.3. *Response Times*

Table 3.5 gives some time constants (i.e., $\tau$'s) for various kinds of thermometers. Anemometers have a similar constant that is discussed later.

**General Errors Associated with Air Temperature Measurements** True air temperature measurements cannot be obtained unless the temperature sensor is in radiative, convective, and conductive equilibrium with its atmospheric environment. This requires that radiation from the sun, especially, earth, and the instrument's surroundings must be minimized, and convection must be maximized (e.g., Middleton and Spilhaus, 1953). Actually there are four basic errors that must be considered in order to obtain the true value of air temperature measurement, namely, transient, windspeed, conduction, and radiation. These errors should be considered regardless of the kind of thermometric sensing device used. Keep in mind that the choice of the recording or readout mechanism may also affect the accuracy. Additional details may be found in Fritschen and Gay (1979), Brock (1984), Middleton and Spilhaus (1953) and Wang (1975). Transient errors are essentially those that arise from the time constant and response characteristics of the sensor and they have been discussed in previous sections of this text.

Windspeed errors, $E_{ws}$, may be expressed (Fritschen and Gay, 1979) as

$$E_{ws} = T_T - T_s = (1 - r)U^2[2gJc]^{-1} \tag{3.8a}$$

where $T_s$ is the sensor temperature, $T_T$ is the "true" air temperature, U is the windspeed, g is the gravitational constant, J is the mechanical equivalent of the thermal energy ($\approx$0.24 cal $J^{-1}$), c is the specific heat of air at constant pressure, and r is the recovery factor. The recovery factor is defined as

$$r = (\text{measured temperature} - \text{static air temperature})(\text{dynamic air temperature})^{-1} \tag{3.8b}$$

where the static temperature is the actual air temperature at all times, the measured temperature is the equilibrium temperature sensed by a stationary ideal-geometry, adiabatic probe, and the dynamic air temperature is the thermal equivalent of the directed kinetic energy of the gas continuum. Be aware that r is proportional to the **Prandtl number** (see Glossary), Pr, as follows: $r = c\,Pr^{0.333}$ for laminar flow, and $r = c'\,Pr^{0.5}$ for turbulent flow (Fritschen and Gay, 1979). c and c′ are proportionality constants which account for the shape, size, and inherent non-ideality and non-adiabatic nature of the probe. Note that Pr is $\approx$0.8 near the surface during turbulent flow, and Pr is $\approx$0.9 near the surface for laminar flow (Fritschen and Gay, 1979).

Friction and adiabatic compression increase the temperature (i.e., dynamic heating), particularly in high windspeeds. Dynamic heating is proportional to $U^2$ and is also influenced by the shape and size of the thermometer. This error should be $\leq$0.02 °C when windspeeds are $\leq$10 m $s^{-1}$, and $\approx$1 °C when the windspeeds are 100 m $s^{-1}$. Consequently, the effect of windspeed can usually be ignored when measuring the air temperature, except in micrometeorological and especially airborne measurements. Note that airplanes typically cruise at $\approx$400 m $s^{-1}$ leading to an error of $\approx$16 °C. The flow past a sensor, or the ventilation of a sensor, is especially important for psychrometers.

The radiation error, $E_r$, is the most common, and potentially the most significant of the errors associated with air temperature measurements. The magnitude of the radiation error depends on the characteristics of the temperature sensor, the intensity of the various kinds of radiation, and the airspeed past the thermometer, among others. The longwave component of the radiation error may be determined by (Fritschen and Gay, 1979)

$$E_r = K_r \sigma_{sb} \varepsilon A_r (T_T^4 - T_w^4)(h A_c)^{-1} \tag{3.9}$$

where $K_r$ is the view factor of the temperature sensor, $\sigma_{sb}$ is the Stefan Boltzmann constant, $\varepsilon$ is the emissivity, $A_r$ is the radiational area, h is the convective heat transfer coefficient, $A_c$ is the area available for convective heat transfer, $T_T$ is the "true" air temperature, and $T_w$ is the temperature of the radiating surface. Shortwave radiation is also a problem but is more difficult to treat. Radiation passes from the sun, clouds, ground, and surrounding objects through the air without significantly altering the surface air temperature. In contrast, a thermometer element, unobstructed from the sun, usually absorbs most of this radiation resulting in an air temperature that differs from the "true" air temperature.

A common way to avoid a radiation error—and not necessarily the best way—has been to put the sensor inside an instrument shelter (Fig. 3.6). The air temperature within these shelters can be ≈2 °F (≈1 °C) higher than the true temperature of the air on a sunny day and ≈1 °F lower than the "true" air temperature on a clear calm night. Large screen shelters of the NWS have a mean ventilation rate of ≈15% of the mean windspeed at 10 m, yielding a ventilation rate of $\leqslant 1$ m s$^{-1}$ for a mean windspeed of 6 m s$^{-1}$ at 10 m compared to the World Meteorological Organization (WMO)'s suggested ventilation rate of 2.5 m s$^{-1}$ for psychrometers (Brock, 1984). This makes a strong case for the use of an Assman psychrometer which has forced ventilation past its temperature sensors.

The basic types of radiation shields are screens, louvered doors, stacked plates, and concentric tubes (including aspirated). Aspirated shields are the standard way to reduce radiation errors. The radiation error may be reduced by

**Fig. 3.6** *Instrument shelter. (Photograph by J. M. Moran.)*

approximately $1/(1+n)$ where n is the number of radiation shields (King, 1943), assuming that the measurements are made in a vacuum with temperature distributions through successive layers of shielding determined by radiation. Convective heat transfer in air augments the radiation exchange and makes the shields more effective, assuming, of course, that the shields are properly spaced. White paint decreases the temperature difference between the cooler surfaces (e.g., water, transpiring crops) and the sensor, but it increases the temperature difference between the hot surfaces (e.g., burned stubble and dry soil) and the sensor (Fritschen and Gay, 1979). Concentric cylindrical shields are very commonly used to shield electronic temperature sensors on commercially available automated and fixed weather towers. The sensor is recessed to a distance at least four times the tube diameter for minimal effects due to radiation. If the sensor is small and highly reflecting in both the visible and infrared then it probably does not need to be shielded from radiation.

An air temperature sensor will not indicate the "true" air temperature if it comes in contact with a surface of different temperature, such as the supports of the temperature device or the electrical leads, due to the conduction of thermal energy. Consider the question, How far should a thermocouple be immersed into a medium to ensure that it is at the temperature of that medium, and is not influenced by a differing air temperature? The answer depends. Suppose a temperature sensor detects a temperature, $T_s$ when placed in a medium with a uniform temperature, $T_\infty$, or the supports at a distance L from the sensor are at $T_\infty$. If $T_s \neq T_\infty$ then a conduction error will occur. The direction of the thermal energy flow depends on whether $T_s$ or $T_{medium}$ is warmer (Fig. 3.7). The conduction error, $E_c$, is simply estimated from the difference between "true" temperature, $T_T$, and the sensor temperature, $T_s$, or as follows (Fritschen and Gay, 1979)

$$E_c = (T_T - T_s) = (T_T - T_\infty)\{\cosh[L(4h)^{0.5}(d\,k)^{-0.5}]\}^{-1}, \tag{3.10}$$

where cosh is the hyperbolic cosine, L is the length of the wire between $T_\infty$ and $T_s$, h is the convective heat transfer coefficient, d is the diameter of the wire or support diameter, and k is the thermal conductivity. Conduction errors can be reduced, by *decreasing:*

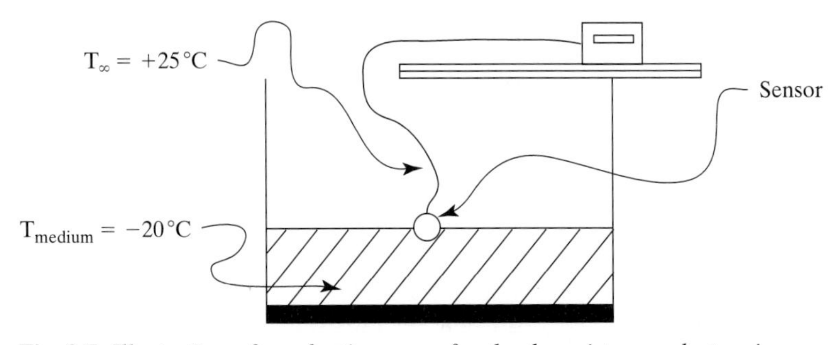


**Fig. 3.7** *Illustration of conduction error for the thermistor or electronic thermometer.*

a. the difference between true temperature and the mount temperature
b. the wire diameter, d
c. the thermal conductivity
   or by *increasing*
d. the convective heat transfer coefficient
e. the junction length, L.

Note that the case of temperature measurements made close to the interface of two different media of different temperatures is more complicated than implied by the discussion of Equation 3.9. A skin boundary layer with steep temperature gradient exists in both media. Furthermore, the sensor body will normally have a different conductivity then either medium, and will thereby disturb the temperature field. A realistic example of a sensor with a thermal gradient is a wet-bulb thermometer. A liquid-in-glass thermometer calibrated for partial immersion, rather than total or complete immersion, should be used to minimize the conduction error on the wetbulb liquid-in-glass thermometer.

However, one should also avoid contaminating a wetbulb sensor with tap water, salt, or other impurities. Simply, the latter can affect electrical conductivity, which could result in an erroneous wet-bulb value and subsequent relative humidity. The presence of salts is a concern in the marine environments and can effect current flow, especially when wet. Remember, a salt is a combination of an equal equivalence or amount of positive and negative ions, meaning that the net charge on the salt molecule is zero. The water frees the ions such that the ions become mobile and will effect any current flow. The freeing of the ions that make up a salt molecule could also result in a local heat release known as heat of hydration (e.g., Atkins, 1986). Salt and other impurities in psychrometer water will cause the wet-bulb to read high (Fritschen and Gay, 1979). Even touching certain sensors with fingers, or holding them too close to your body can have an adverse effect on the measurement of temperature and humidity.

In summary, we have learned the fundamentals of making temperature measurements using a number of temperature-measuring devices on a couple of different platforms, such as ground-based and satellite-borne platforms. We then considered the response time and time constants of various thermometers and some of the errors that might effect their indication of the true value of the air temperature.

Caution is raised over the estimation of air temperature at any level in the atmosphere using remote-sensing devices, especially those on satellite platforms. The values obtained from such platforms rely heavily on computer models that are "relatively" young in development. That is, the deconvolution of air temperatures from satellite measurements are essentially preliminary, although the results are quite promising. Caution is especially prudent when reviewing the vast number of global climatic datasets and the recent satellite-based climatologies of surface and atmospheric temperatures. For example, are the surface air temperatures over the ocean adequately represented? The quality of the air temperature measurements and the density of the reporting stations are minimal and they certainly have not stayed the same over the past 100 years or so. This does not mean that one should dismiss the satellite temperature values or the temperature values obtained from other sensors. One does need to realize the procedures used to obtain the measurements and the inherent consequences of such before applying the data to the particular need(s). One way to ensure that air and surface temperature retrievals are improved or adequate is to increase the number of studies that relate air and surface temperatures to those derived from satellite and ground-based remote temperature-sensing devices for many different meteorological and surface conditions.

## *Exercises*

1. A thermocouple has a Seebeck coefficient, N = 40 $\mu$Volts/°C (reference to ice). The Seebeck coefficient is not constant with temperature. It increases with increasing temperature. Determine the values indicated by the question marks in Fig. 3.8. Show your work.

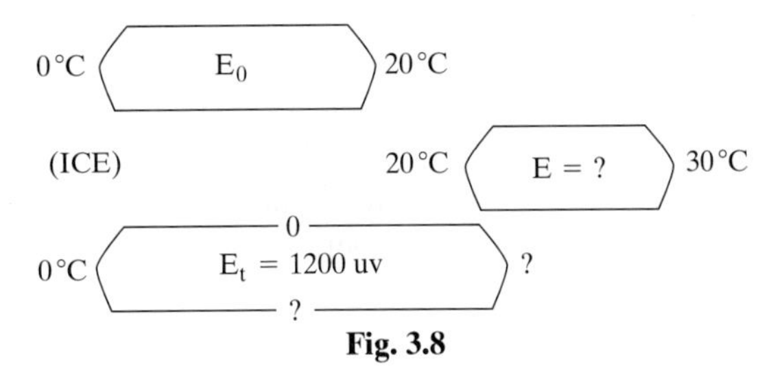


**Fig. 3.8**

2. A. Using a thermocouple and a strip chart recorder, determine their time constants in a water bath. (Make three estimates of the time constant. Make one estimate using the slope of the response curve).
   B. Measure the resistance of a thermistor at 0, 25, 100 °C. Plot LOG R vs. $T^{-1}$. Determine beta from this plot.
   C. Design a bridge circuit that allows for the thermistor in (B.) to very accurately make measurements in the range 0 to 50 °C. What is the full scale output and maximum self-heating?
   D. Using a standard thermocouple, infrared thermocouple, and a thermometer make the following thermocouple output measurements using a potentiometer.
   a. Unknown vs room temperature
   b. Unknown vs ice bath
   c. Room temperature vs ice bath
   Show that these measurements are consistent with the Law of Intermediate Temperatures.

3. Suppose you just received your first thermometer. You have to try it out immediately. Consequently you bring the thermometer outdoors. The indoor temperature was 30 °C, and the outdoor temperature was 20 °C and does not vary during your measurements. The thermometer has a 0.1 °C resolution, and its output is related to a step input. How long will it take for the thermometer to reach equilibrium with the outside air? What is its time constant?

4. Given that the change in the sensor's temperature proceeds at a constant rate of $\beta$ degrees per sec (sometimes termed a "RAMP" input): (a) derive the expression for the

time constant; (b) determine the magnitude of the time constant; and (c) determine the time it will take for this thermometer to come into equilibrium with the outside air. Use the values provided in Exercise 3.

5. Given the same conditions in Exercise 3, except that the input is sinusoidal: (a) derive the expression for the time constant; (b) determine the magnitude of the time constant; (c) determine the time it will take for this thermometer to come into equilibrium with the outside air; and (d) figure out how the time constant of this thermometer compares with that in Exercises 3 and 4?

6. A radiosonde ascent is made during the morning hours of a day when the environmental lapse rate, $\Gamma$, is linear and $= 10\,^{\circ}\mathrm{C\,km^{-1}}$, and the surface temperature is $T_0 = 20\,^{\circ}\mathrm{C}$. Let $T_e$ denote the environmental temperature. The balloon ascends at a rate w, and has an electronic temperature sensor to measure the temperature during the ascent. (a) Derive the expression for the time constant; (b) determine the magnitude of the time constant; and (c) determine the time it will take for this thermometer to come into equilibrium with the environmental air. [You may want to reread Chapter 1, Section 1.1.1. Assume that the temperature sensor is flat and rectangular with area $A_{th}$, volume, $V_{th}$, and a heat capacity per unit volume, $C_{th}$, $w = 6.0\ \mathrm{m\,s^{-1}}$ and is constant with height.]

7. Make air temperature measurements at various times, t, $= 0, 0.5, 1, 2, 5, 10,$ and 15 minutes with a mercury thermometer and an electronic thermometer. Repeat at a second location. [Note: Keep the thermometer covered until measurements begin, write down the temperature on the thermometer as soon as you take it out of its cover, as well as the time. Then note subsequent temperatures for the various times. Be careful not to introduce any systematic or random errors.] Locations should include one or more of the following;
   a. GRASS–SHADE$^{+}$
   b. GRASS–SUN$^{+}$
   c. ROOF–SHADE$^{\circ}$
   d. ROOF–SUN$^{\circ}$
   e. CONCRETE–SHADE*
   f. CONCRETE–SUN*
   g. OTHER

   $^{+}$Repeat the same procedure for the temperature of the grass.
   $^{\circ}$Repeat the same procedure for the temperature of the roof.
   *Repeat the same procedure for the temperature of the structure.

   Plot temperature versus time for each set of data.
   Estimate the time constant using: (a) the slope of the response curve, and (b) the relevant theory. Discuss your findings, including all possible errors.
   Hint: A graphical determination of the time constant should be done on semi-log paper, and the values should form an approximately straight line since the measurements were made with a first order measuring device. Of course, if no semi-log paper is available, then simply calculate the log base 10 of your y variable and plot on a linear scale.

CHAPTER FOUR

# Pressure Measurement

This chapter concentrates on the two most common instruments used to make atmospheric pressure measurements, namely the mercury Fortin-type barometer and the aneroid barometer. In particular, it describes the basic principles by which they make their measurements and then discusses the errors associated with pressure measurements. There is also a brief discussion of some other pressure-measuring devices at the end of the chapter.

## *4.1. Pressure Measurement Principles*

Pressure, P, is an omni-directional force, F, exerted onto a unit area, A, or $P = F/A$. Atmospheric pressure units are commonly expressed in inches or millimeters of mercury (Hg), kiloPascals (kPa), or millibars (mb). The most relevant kinds of pressure we may need to measure in meteorology are air (or atmospheric) and water vapor. The pressure of a gas, such as air and water vapor, may be defined by the equation of state,

$$P = \rho R T, \tag{4.1}$$

where $\rho$ determines the number of molecules that hit a wall, T is a measure of the velocity of these molecules (i.e., air temperature), and R is the universal gas constant. Thus, gas pressure (bombardment of molecules against each other and the walls) is due to molecular motions. Pressure is actually part of a stress tensor, $\tau_{stress}$, of which there are 9 components (Fig. 4.1). The components perpendicular to each plane of a three-dimensional Cartesian coordinate system, for example, (namely, $\tau_{stress;xx}$, $\tau_{stress;yy}$, and $\tau_{stress;zz}$) are the pressure in the x,y, z directions respectively, and the others are stresses.

Atmospheric pressure may be mathematically expressed via the barometric height formula and in simplified form (assuming the hydrostatic approximation is valid) as

$$\partial P = P - P_{sealevel} = -\rho g(Z - 0) = -\rho g Z, \tag{4.2}$$

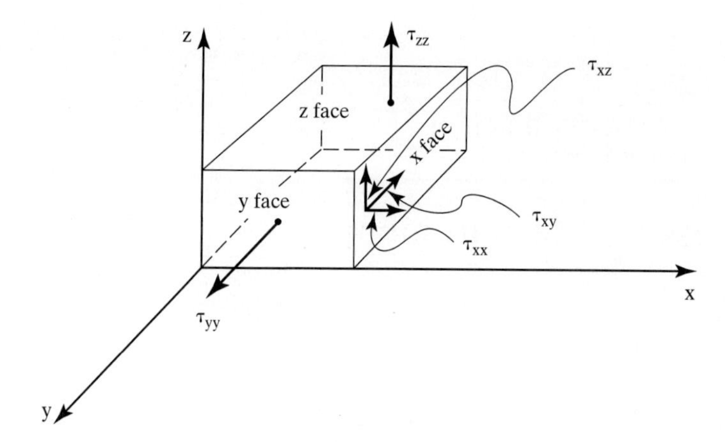


**Fig. 4.1** *The nine components of a stress tensor. The orthogonal components describe the pressure.*

where Z is altitude, g is gravity, and $\rho$ as previously defined. The measurement of air pressure depends on

1. temperature,
2. gravity,
3. height relative to a reference level (sea level).

The following outlines the importance of items 1 through 3 with respect to the determination of pressure and are also important for correcting measured pressure readings to sea level.

1. The temperature dependence or correction, $\Delta P_T$, is very important because of the coefficient of expansion of the glass, mercury, and the brass scale of the barometer. It is derived from the difference between the average air temperature and the air temperature of a standard atmosphere in an imaginary column with unit cross-sectional area that extends upward from our barometer at the surface. The temperature dependence or correction, $\Delta P_T$, relative to a reference temperature of 0 °C is obtained (e.g., U.S. Department of Commerce, National Weather Service, 1963; Letetsu, 1966; List, 1971) using

$$\Delta P_T = P_r\{([1 + (1.84 \times 10^{-5}\,m(m\,°C)^{-1}T)] \div [1 + (1.818 \times 10^{-4}\,m^3(m^3\,°C)^{-1}T)]) - 1\} \quad (4.3)$$

where $P_r$ is the uncorrected pressure reading and T is temperature.

2. Gravity must be accurately known in order to determine exact pressure differences. Gravity changes with latitude as well as altitude. The gravitational force between a mass of air above the surface of the earth and the mass of the earth is inversely related to the separation between the center of masses of the earth and the air squared. Thus, the force of gravity should increase toward the pole, since the dis-

tance at the pole to the center of earth's gravity is closer by $\approx$300 km than at the equator. Actually, if we assume the earth is a geoid, then the gravity at sea level would be 9.7805 m s$^{-2}$ at the equator, 9.8063 m s$^{-2}$ at 45° north or south latitude, and 9.9322 m s$^{-2}$ at the poles. Anomalies are generally within $\pm$0.0002 m s$^{-2}$ of gravity, but they may have a value on the order of a magnitude less than 0.0002 m s$^{-2}$ in field measurements. The latitudinal variation of gravity is usually neglected for atmospheric pressure measurement, and the gravity dependence or correction, $\Delta P_g$, may be expressed as

$$\Delta P_g = P_T\{([980.616/980.665][1 - (2.6373 \times 10^{-3}\cos(2\phi)) + (5.9 \times 10^{-6}\cos^2(2\phi))]) - 1\} \tag{4.4}$$

where $P_T$ is the temperature-corrected pressure reading and $\phi$ is the latitude in degrees north or south.

3. The height dependence is primarily important when adjusting your station pressure to sea level, and may be expressed as (e.g., U.S. Department of Commerce, National Weather Service, 1963)

$$\Delta P_S = P_0\{1 - [1 - ((6.5 \times 10^{-3}/288.16)Z)]^{5.2561}\}, \tag{4.5}$$

where $P_0$ is the standard pressure at sea level and Z is the station elevation above sea level in meters.

The barometer used for obtaining atmospheric air pressure is either one of two general types, namely, cistern and non-cistern. This chapter concentrates on the cistern barometers which include the liquid-in-tube (indirect measure of air pressure) and fluidless flexible tube (a relatively more direct measure of air pressure) barometers. The liquid-in-tube barometer uses the hydrostatic approximation to obtain changes in pressure from changes in the height of the fluid in tubes of known circumferal area. Mercury is the most commonly used fluid, despite its poisonous nature. Water is not used to obtain the air pressure because it is 14 times less dense than mercury and more prone to evaporation. A water barometer (sometimes termed a manometer) would also require a tube that is 10 m (32.8 ft) in length in order to cover the typical ranges of atmospheric air pressure. The mercury Fortin-type barometer (Fig. 4.2a) is an example of this category of barometers, and operates according to the principle of the Cistern (with fluid) or Torricelli barometer (1 Tor = 1mm Hg):

$$P = \rho_{hg} g h_{T_0},$$

where $h_{T_0}$ is the height of the mercury column at a standard temperature and elevation. Figure 4.2b–d shows the scales, vernier, and mercury column height adjusting screw, respectively, used to indicate the air pressure prior to its correction for temperature, gravity, and height relative to sea level. Further details on the mercury Fortin-type barometer may be found in Middleton and Spilhaus (1953), and BMOAM (1961), for example. Exercise 3 at the end of this chapter provides the

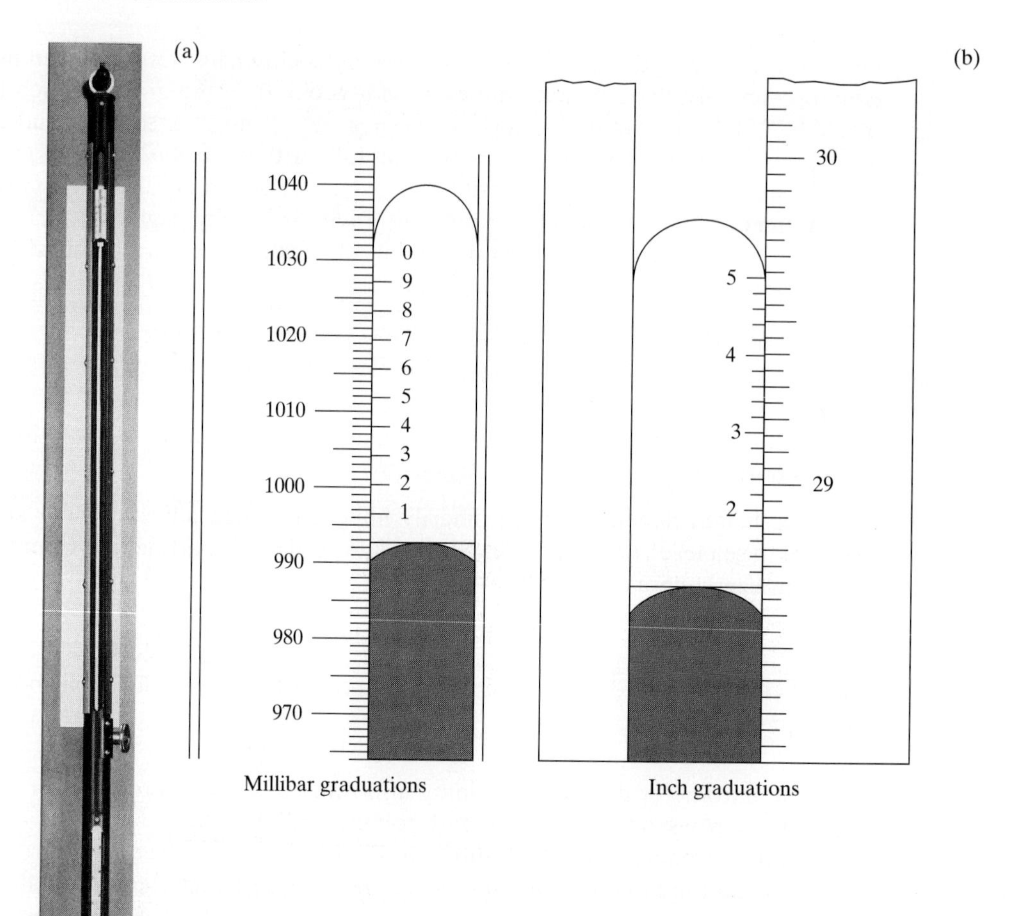


**Fig. 4.2** *(a) Photo of the mercury Fortin barometer. (Courtesy of G. Pfaff of Princo Instruments, Inc.) (b) Scale, (c) vernier, and (d) cistern areas of the mercury Fortin barometer. (Adapted from BMOAM, 1961.)*

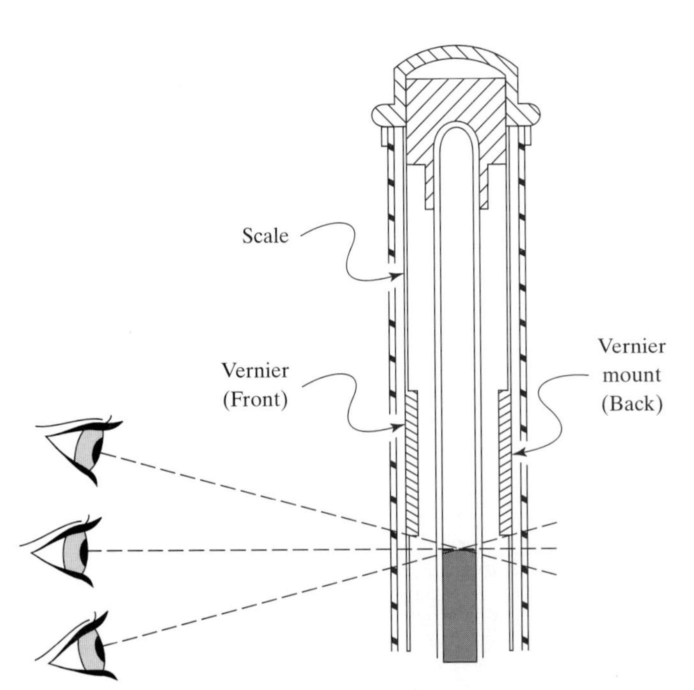


**Fig. 4.2** *(continued)*

procedure for converting the uncorrected pressure reading of a mercury Fortin barometer into sea-level pressure.

The fluidless barometer consists of three basic configurations, namely the aneroid barometer, hypsometer, and the bourdon tube. The aneroid barometers are most commonly used, although the bourdon barometer is generally used in high temperature industrial situations. The aneroid barometer (Fig. 4.3) is a thin metal membrane or diaphragm that deforms in response to external pressure. Usually two membranes form the walls of a closed chamber with one of them fixed. It operates on the principle symbolized by the hydrostatic equation, equation of state, and the hypsometric equation. For example, start with the hydrostatic equation, Equation 4.2, and replace density by the equation of state,

$$\partial P = -Pg(RT)^{-1}\partial Z.$$

Assume T is independent of P and Z, simplify, and integrate.

$$\int_{P_0}^{P}(\partial P P^{-1}) = -\int_{Z_0}^{Z}(g(RT)^{-1})\partial Z \rightarrow \ln(P/P_0) = -(g(RT)^{-1})\int_{Z_0}^{Z}\partial Z \tag{4.6}$$
$$P - P_0 = \exp[-(g(RT)^{-1})(Z - Z_0)] = \exp[-(Z - Z_0)H^{-1}]$$

where H is the scale height = $RT_vg^{-1}$, and $T_v$ is the average virtual temperature of the layer bounded by P and $P_0$. The aneroid barometers are generally not as accurate as the mercury barometers, but they are liquid-free, portable, lightweight, easily

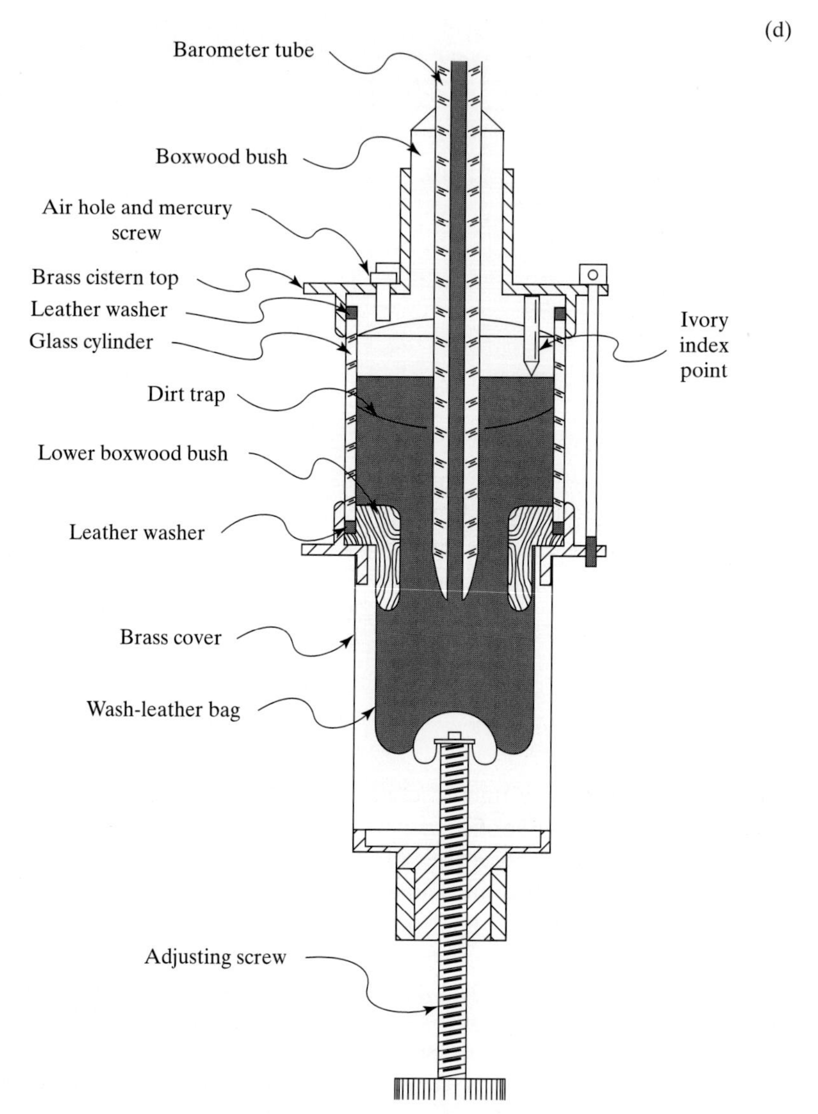


**Fig. 4.2** *(continued)*

adapted to a recording device, and are not generally effected much by movement or acceleration. They must still be calibrated!

## 4.2. *Pressure Measurement Errors*

All barometers are prone to errors caused by (a) wind, (b) uncertainty in temperature of the instrument, and (c) motion of the instrument. When moving air encounters an obstruction it generally causes a change in total atmospheric pressure at the surface of the obstruction due to the change in air velocity and is usually called

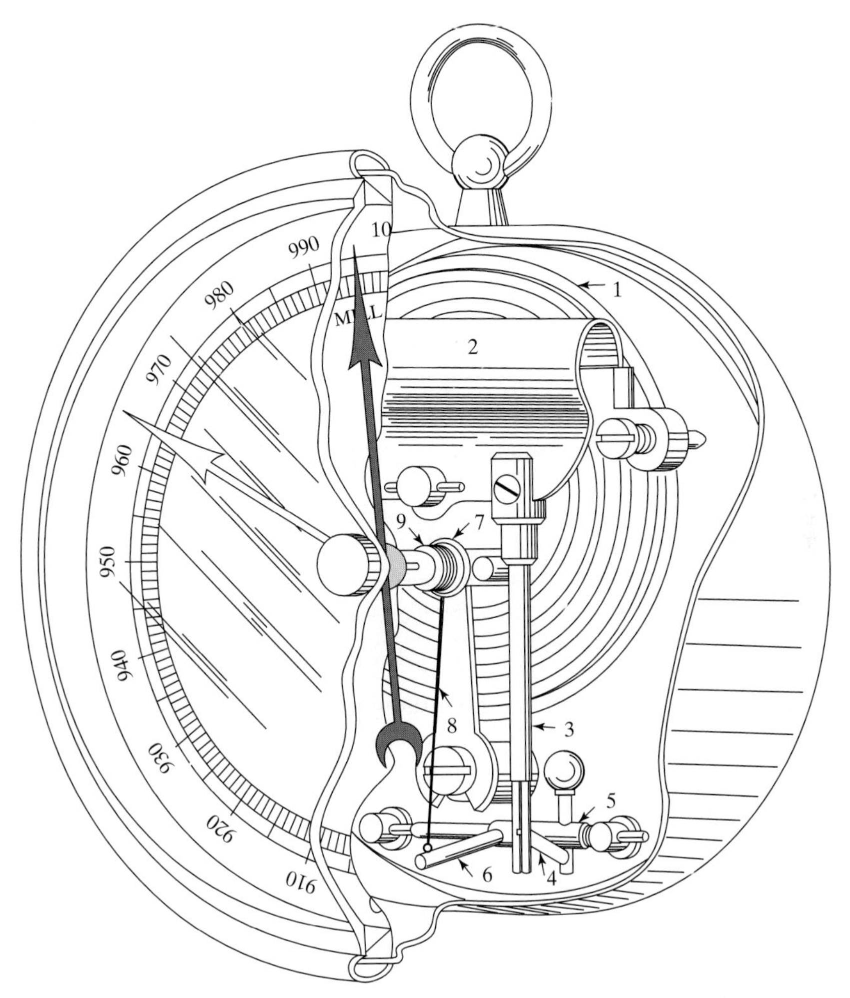


**Fig. 4.3** *Aneroid barometer. 1 is the vacuum chamber, 2 is the spring, 3 is an arm, 4 is a connecting link, 5 is the rocking bar, 6 is the projecting arm, 7 is the hair-spring, 8 is the arbor chain, 9 is the pulley. (Adapted from BMOAM, 1961.)*

dynamic pressure. Note that dynamic pressure changes rapidly near a ridge due to acceleration and deacceleration of the wind (or crowding and spreading of streamlines). This effect is also noticeable at the windward and leeward side of a house, even if the barometer is not directly exposed to the wind. Assuming that the air is completely stopped, the resulting pressure is equal to $0.5\,\rho V^2$, where V is the air speed and $\rho$ is the air density. Consider Example 4.1.

---

**Example 4.1**

What would the indicated pressure of the wind be, if it is measured on a surface normal to the direction of the wind flowing at 40 knots? The pressure is equal to $\approx$2 mb. [*Hint*: use $0.5\,\rho V^2$, and watch your conversion factors)].

---

The pressure in our example is large and one should make sure that our air pressure measurement is of the "air" and not of the wind. The magnitude of the latter error depends on the nature and position of the openings within the room (namely, doors, windows, chimneys, heat vents, etc.) in relation to the direction of the wind. The barometer should be mounted on the interior side of an interior wall that is in an isothermal, temperature-controlled, gust-free room, and out of direct sunlight or other heat source and away from a door. The mercury within the barometer should ideally be isothermal, and corrections of $\leqslant$0.03 mb (assuming a steady rise in external temperature of 2 °F $h^{-1}$), are applied in many instruments in order to ensure this (BMOAM, 1961). The errors in the thermometer reading are usually negligible compared with those due to the uneven temperature of the barometer. Most barometers, especially the liquid type, are not designed to be moved. Both mercury Fortin and aneroid barometer corrections should be redetermined if these instruments are moved. Additional significant errors are briefly discussed for the two general kinds of barometers as follows.

***4.2.1. Errors in Mercury Fortin Barometer Measurements.*** The residual pressure in the vacuum space of a Fortin barometer should not exceed the amount corresponding to the reading error associated with the instrument under the best conditions. A defective vacuum is usually caused by the presence of water vapor or air. The presence of air may be detected by tilting the barometer as if to fill it, and noting whether a bubble is seen. The general rule of thumb is: a $\leqslant$1 mm diameter bubble when the tube is laid flat is tolerated for every 6 mm bore tube. For example, a tube with a 12 mm bore containing a >2 mm diameter bubble would not be tolerated. The error due to water vapor ($H_20_v$) is more difficult to determine since it only takes a small amount to affect the readings. For example, 0.01 mg of water vapor would cause an error of $\approx$2.3 mb at normal pressures. Nonetheless, the presence of air or water can be shown by comparing the actual reading of the barometer with a standard over a range of pressures at a given temperature. The errors will increase as the vacuum space decreases in accordance with Boyle's law.

Errors due to the barometer tube not being vertically mounted are usually small but are not always negligible. The error caused by inclining the barometer tube from the vertical will vary according to the direction in which the incline takes place (Fig. 4.4), and its magnitude varies between f and j, where

$$f = \{(h/\cos\theta') + (d\tan\theta') - h\} \text{ and}$$
$$j = \{(h/\cos\theta') - (d\tan\theta') - h\}, \qquad (4.7)$$

$\theta'$ is the inclination from the vertical, h is the true height of the mercury column, and d is the distance between the center of the mercury column and the half-way point of the radius of the cistern.

---

**Example 4.2**

If $\theta'$ = 5 min. of arc and d = 0.5", then what is the maximum error in a Fortin-type barometer due to the given deviation from the vertical? The maximum error is $\approx$0.026 mb for a pressure of 1000 mb. Thus the lowest part of the barometer must be kept to within 1.5 mm (0.06") of vertical in order to keep the error of non-verticality to within 0.026 mb.

---

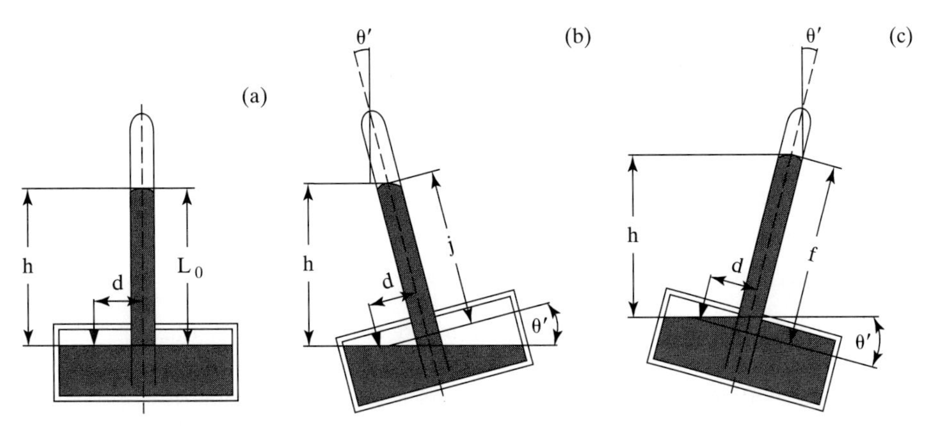


**Fig. 4.4** *Error caused when a mercury barometer is not mounted vertically. (Adapted BMOAM, 1961.)*

The incomplete compensation for the amount of mercury in a barometer or the existence of an improper amount of mercury in the barometer is called the *capacity error*. The capacity error is usually minimal for a new instrument. This error may become important in repaired barometers particularly since it is the glass tube that usually requires replacing and the tube might not have the same bore diameter as the original. The effect of changes in the internal diameter of the upper portion of the barometer tube, in particular, may cause errors in capacity by as much as 0.31%, or an error in the pressure readings of $\approx$0.31 mb/100 mb of the scale. There is also the possibility that this error may vary throughout the barometer tube within the working range of the instrument due to (1) the lack of uniformity of the bore of the tube or cistern, and (2) the mercury level in the cistern taking up a position beyond the uniform cylindrical portion. The lack of bore uniformity is not usually significant, whereas the mercury level beyond the uniform cylindrical portion may occur in two ways. The mercury may reach the cistern top before the lower working limit is reached, or else it may reach a flange or shoulder in the middle of the cistern before the upper working limit is reached. Flanges are not normally present in some of the older barometers, and this error is not usually met in modern mercury barometers.

An index error will occur if any imperfections exist in the construction and adjustment of the barometer. It includes errors from residual gas in the vacuum space, refraction in the glass tube, or an improper amount of mercury in the barometer. These are determined by comparing instruments with a standard whose index errors are precisely known. Personal errors due to improperly reading the mercury meniscus can cause errors of the order of 0.002" (0.07 mb). These errors (excluding the personal component) are normally determined by the manufacturer or other standardized lab, and are commonly reported to the nearest 0.1 mb with an accuracy of $\pm$0.2 mb at several points in the range. The accuracy of a pressure reading is $\pm$0.01 mb = $\pm$1 foot (ft) at normal atmospheric pressure and temperature. A higher sensitivity is required at higher altitudes (i.e., if on Mt. Everest). It is particularly noteworthy to state that an error in pressure by $\pm$0.01 mb corresponds to a temperature measurement accuracy of $\pm$1 °C.

There are a few additional errors unique to marine mercury barometers. Namely, those that arise from lag, the motion of the barometer, the motion of the ship, the presence of the constriction in the barometer tube, and vibration. A detailed discussion of marine mercury barometer readings and their errors is provided in BMOAM (1961), and only the errors are briefly presented.

Lag errors are determined in a very similar fashion to those for thermometers, and are governed by the following equation

$$(\mathrm{dh\,dt^{-1}}) + [(\mathrm{L^{-1}(h - h_0)})] = 0 \tag{4.8}$$

where h is the true height of the mercury above the cistern level, t is time, L is the response time and is a constant for a given barometer, and $h_0$ is the corrected height of the mercury above the cistern. When the pressure is changing at a constant rate the barometer reading lags behind the "true" reading by an amount equal to $(\alpha L)$ where $\alpha$ is the rate of change of pressure (i.e., pressure tendency), provided that the time interval since the change began is large compared with L. Given L is $\approx$6 to 9 min. when the pressure changes at a rate of 1 mb $h^{-1}$, then the error at any time is $\approx \pm 0.15$ mb. (The barometer reads high if the pressure is falling and low if the pressure is rising.) Note that if this rate of change was constant it would be possible to eliminate this error by reading the barometer at a time L min. later than that for which the pressure is required.

There is an additional error that must be accounted for if one uses a mercury barometer on an ocean vessel, namely errors due to the swaying of the vessel. The errors due to a swaying barometer manifest themselves in one of two ways, namely: (1) an increase in the mean length of the mercury column due to the inclined position of the barometer to the vertical, (2) a decrease in the length of the column because of an apparent increase in the value of the gravitational acceleration due to the inertial force set up by the swinging barometer. It may be shown that these errors can be combined in the correction term of the form

$$0.25\,\mathrm{h}\,\theta'^2\{1 - [4\pi^2(\mathrm{h_1 - h_2})(\mathrm{g\,t^2})^{-1}]\}, \tag{4.9}$$

where $\theta'$ is the angle which the barometer makes with the vertical at the extremity of the swing, h is the true mercury column height, t is the time of oscillation of the barometer, and $h_1$ and $h_2$ are the distances below and above the axis of swing of the surfaces of the mercury in the cistern and the tube, respectively. This error is minimized if the natural period of swing and the mean barometric height are assumed and the axis adjusted accordingly. A rough estimate of the magnitude of this error is illustrated by Example 4.3.

---

**Example 4.3**

If $h_1 = 53$ cm, $h_2 = 23$ cm, $t = 1.5$ s, $p = 1013$ mb, and $\theta' = 5°$ (0.087 rad.), then what is the magnitude of the correction due to swaying? It is +0.9 mb and the barometer would be reading too high in this case.

---

The full magnitude of this error will be realized if the oscillation occurs for at least 15 to 20 min. It is also necessary to note that this discussion does not refer to a fixed axis relative to the ship.

Errors due to the movement of the ship are from the increase or decrease in the inertial force induced by the earth's rotation acting on the ship as it proceeds with an east–west component. This changes the acceleration due to gravity by the amount $-2\,\omega \mathrm{v} \cos\phi \sin\theta$, where $\omega$ is the angular velocity of the earth, v is the velocity of the ship, $\phi$ is the latitude of the ship, and $\theta$ is the direction of the ship's course determined relative to true north. The correction is 0.048 cm $s^{-2}$ if v = 10 kt., $\phi = 50°$, and $\theta = 90°$. The corresponding correction to the barometer reading at 1000 mb is 0.5 mb. The barometer indicates a greater than "true" value pressure when the ship is moving eastward and smaller than "true" value when it is moving westward. This error is usually neglected compared to the other errors that might occur in measuring pressure on a ship, but it should not be ignored. Rolling, pitching, pumping, yawing, and vibration make it very difficult to read the meniscus of the barometer. The error due to windspeed is greater on ships than most land platforms primarily because of the higher windspeeds usually encountered and the less than ideal placement of the barometer at sea (recall Example 4.1). Although it is difficult to assign a quantitative estimate of the errors of shipboard pressure measurements, it seems very likely that the errors in the mean pressure are of the order of $\approx$1 mb in moderate winds, and they are larger in more severe conditions. The difficulty in reading a swaying, or vertically moving, barometer that is also subject to vibration introduces additional uncertainty that can not be determined.

***4.2.2. Errors in Aneroid Barometer Measurements.*** Errors associated with the aneroid barometers are primarily due to temperature changes, hysteresis, and improper reading. The error due to changes in temperature may be approached in at least one of two ways that are detailed elsewhere (e.g., BMOAM, 1961; the US Department of Commerce, National Weather Service (Bureau), or equivalents in other countries). Simply, if $p_0$ is a reference pressure, and $\partial T$ represents the positive change in temperature in the vicinity of the barometer, then the resulting change in pressure $\partial p$ may be expressed as

$$\partial p = \partial T\left[(p_0\, 273^{-1}) + ((p - p_0)\, b)\right] \tag{4.10}$$

where p is the atmospheric pressure and b is a constant dependent on the material (e.g., b = −0.00026 for steel). Hysteresis is a non-uniform aneroid deflection, relative to an initial pressure reading, that occurs when the aneroid is subjected to a cyclic pressure variation. The width of the deflection ultimately reaches a steady dimension (Fig 4.5). A related deflection, known as "creep" occurs when the pressure is constant (Fig 4.5). A continual, but gradual change in the magnitude of the hysteresis or related deflection is most likely due to small changes in the structure of the chamber, and is termed secular change. The hysteresis and related errors (a) essentially increase when the temperature increases, and (b) are more important in aircraft instruments than they are in surface aneroids. The increased temperature

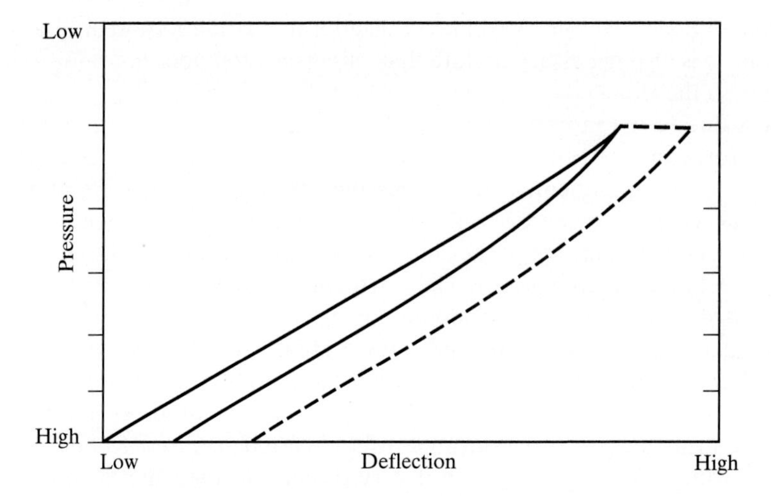


**Fig. 4.5** *Hysteresis effect. Dashed line represents creep. (Adapted from BMOAM, 1961.)*

increases the range of pressure variation, and the rate of pressure change. The BMOAM (1961) is an excellent source for additional details of the previous two barometers and the errors associated with their measurement.

## 4.3. *Other Pressure-Measuring Devices*

There are additional instruments that may be used to measure pressure, such as the Pitot tube and altimeter. The Pitot tube is simply a piece of tubing that is typically bent at right angles with one concentric longitudinal axis aligned with the fluid flow and the other connected to a manometer (Fig. 4.6). It indicates windspeed by detecting a difference in the manometer fluid level, known as the manometer deflection, $\Delta Z$, and operates according to Bernoulli's principle (e.g., Yuan, 1967)

$$P + (0.5\rho U^2) + (\gamma \Delta Z) = \text{constant}, \tag{4.11}$$

where U is the mean windspeed, P is pressure exerted by the fluid (air in this case), $\rho$ is the density of the fluid (air in this case), and $\gamma$ is the specific weight of the manometer fluid. The pitot tube is primarily used on aircraft. The altimeter is simply an aneroid barometer calibrated in $\Delta Z$ with a temperature dependent scale. That is they measure the height relative to a reference level (i.e., sea level), and then use Equation 4.2 to obtain a pressure value.

The fundamentals of modern air pressure measuring devices and the errors associated with those measurements have been discussed for a mercury Fortin barometer, aneroid barometer, and to a lesser extent the pitot tube and altimeter.

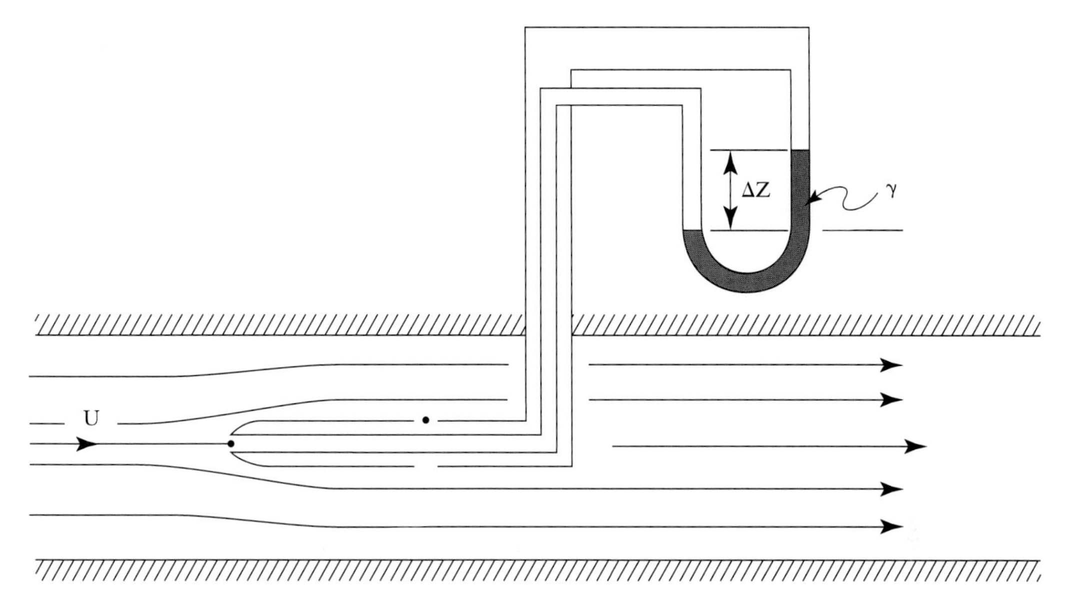


**Fig. 4.6** *A Pitot tube. (Adapted from Yuan, 1967.)*

One of the exercises below will simply provide you with the procedure for reading a mercury Fortin barometer before it asks you to convert the reading into a sea-level pressure.

## *Exercises*

1. An altimeter is set to the correct reading at the beginning of one's journey up a hill that extends from sea level to an altitude of exactly 2 km over a 6 hour period. Meanwhile, the pressure drops 15 mb during the same 6 hour period due to an approaching front. What is the altimeter reading at the top of the hill?
2. A lake-breeze is common in the afternoon, when the land surface becomes warmer than that of the nearby lake. A pressure gradient is set up at all altitudes where the cool air moves landward, and the top of the lake-breeze is the altitude where the pressure gradient is zero. What is the approximate height of the top of the lake-breeze, when the surface pressure is 1 mb greater over the lake than over the land, and the temperature difference is 4 °C? Assume that the surface pressure is 1000 mb, and the surface temperature over the land is 10 °C. Furthermore, assume that the temperature profiles over the land and water are isothermal.
3. The following procedure is commonly used to convert the reading of a Fortin barometer into sea-level pressure.
   a. Read the attached thermometer and note the temperature.
   b. Set the adjusting screw at the bottom of the cistern so that the mercury surface just comes in contact with the ivory reference point located on the right side at the top of the cistern. Actually, this point should just touch its reflection on the mercury without disturbing the mercury surface.

c. Gently tap the side of the barometer tube at the level of the upper mercury surface to allow the meniscus to adjust to the new mercury level.
d. Then set the vernier scale (Fig. 4.2b) so that its bottom appears to just touch the top of the mercury surface. Your eye should be at the level of the top of the mercury column when you do this, and the front and rear edges of the scale should align with the top of the mercury.
e. Lower the mercury level in the cistern to slightly below the ivory reference point. Do not change the scale adjustment, and do not lower the mercury more than two mm!
f. Read the scale to the nearest 0.001′ or 0.1 mm of mercury.
g. Note the time, and make sure the temperature indicated on the attached thermometer has not varied since you made your measurement.
h. Apply a temperature correction and gravity corrections (i.e., see Equations 4.3 and 4.4).
i. Apply the sea-level height differential to convert the station pressure in h. to sea-level pressure, i.e., see Equation 4.5.

Suppose the pressure you read is 754.2 mm and the attached thermometer is 24 °C at a station of 700′ elevation above sea-level. What is the sea-level pressure?

CHAPTER FIVE

# Windfield Measurement

This chapter covers the fundamentals of windfield measurements made by traditional and non-traditional measuring devices. The traditional devices are those which are commonly used today to measure the windfield, such as the cup and hot-wire anemometers and the wind vane. In contrast, the so-called non-traditional wind-measuring devices include the sonic, laser, and microwave anemometers, as well as the radar wind profiler. Each device is described with respect to its basic operation principles, time constants, and accuracies. There is also a section that addresses the deployment of anemometers in the field.

The science of measuring and recording the windfield is called *anemometry.* The term *windfield* is used to combine both the speed and direction of air in a space–time continuum. The rate of air movement in its instantaneous direction is termed the wind velocity. Wind velocity is a vector, whereas windspeed is a scalar quantity. Windspeed is usually specified in knots, kt, or meters per second, m $s^{-1}$ (note that 1 kt = 0.515 m $s^{-1}$ = 1.152 mi $h.^{-1}$). The direction of the wind is specified relative to true north at the place of observation and refers to the direction from which the wind is blowing in units of degrees (clockwise from true north) or compass point. Surface windspeed measurements have long been made with instruments known as propeller anemometers, of which the gill and cup anemometers are examples. Surface wind direction has long been measured by wind vanes. The most recent and futuristic wind devices are and will be able to simultaneously monitor the windspeed, wind direction, and wind perturbation (i.e., wind gusts and lulls) from any platform (e.g., Baker et al., 1995; Atlas et al., 1996).

Anemometric devices have been categorized in a number of ways, e.g., anemometers, vanes, combination anemometers and vanes (Figs. 5.1 and 5.2), remote wind sensors, pitot tubes, and pilot balloon tracking, or aerodynamic, thermodynamic, and frequency shift anemometers. The remote sensing category, so termed for convenience, contains devices that are unique in design and operation compared to the other categories and include the Doppler radar, satellites (Atlas et al., 1996), Global Positioning System, (see Chapter 8, section 8.1) and lidar (Baker et al., 1995). The pitot tube, pressure plate, pendulum, cup and gill anemometers, and, of course, the wind sock, are examples of aerodynamic anemometers. The pressure plate and pendulum

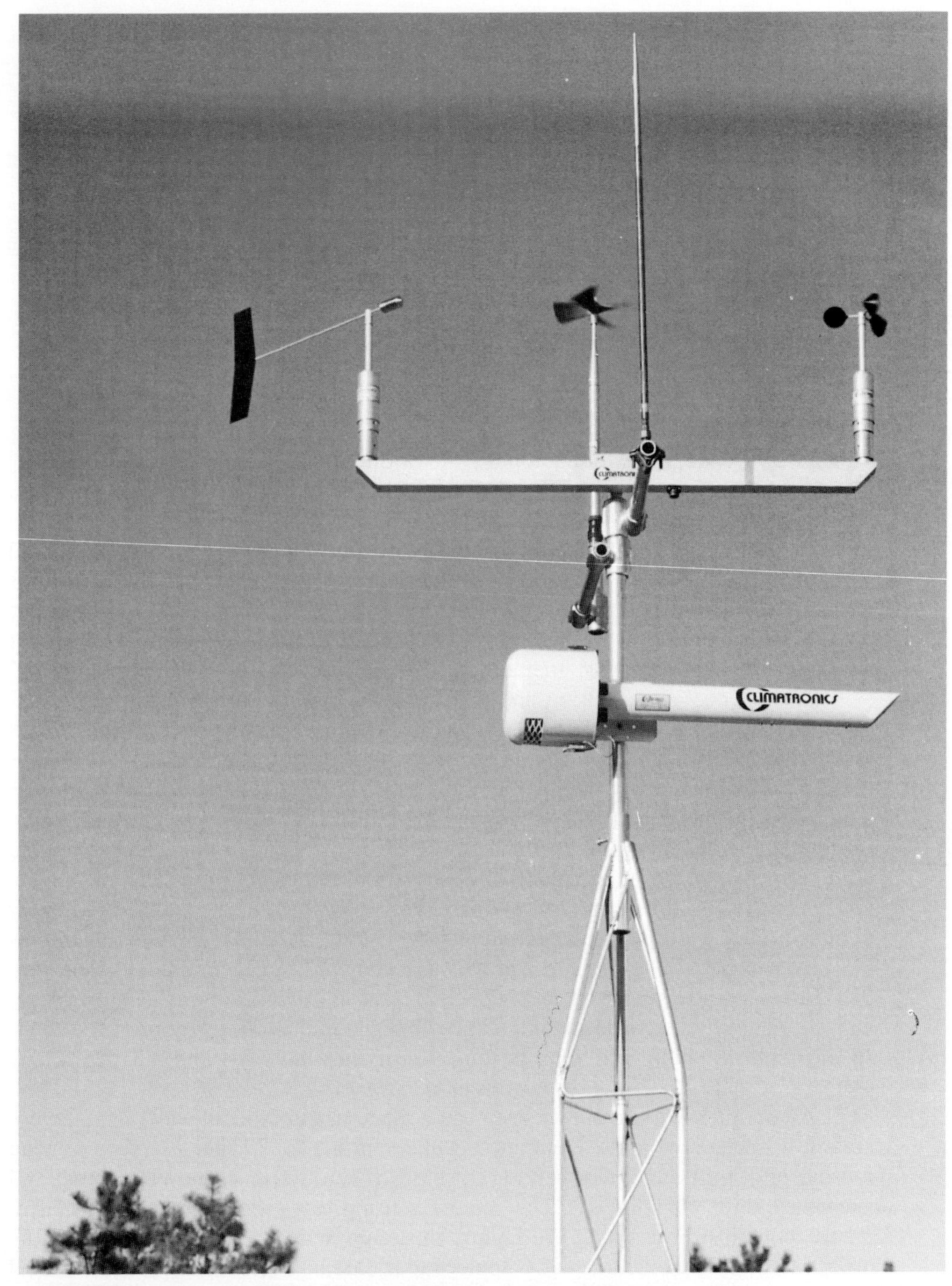


**Fig. 5.1** *Photo of a wind vane and cup anemometer, with a w component gill anemometer and an aspirated shield covering a temperature sensor. (Courtesy of D. Gilmore and D. Katz of Climatronics Corp.)*

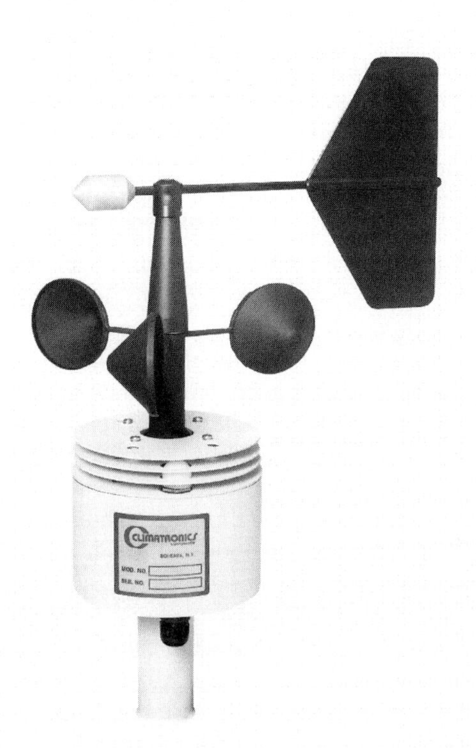


**Fig. 5.2** *A photo of a stacked anemometer vane assembly. (Courtesy of D. Gilmore and D. Katz of Climatronics Corp.)*

anemometers are ideal for elementary to middle school science fair projects or science classes. The British Handbook of Meteorological Instrumention (BMOAM, 1961) states that Hooke (of spring constant fame in physics) introduced the pendulum anemometer to Great Britain sometime before 1667, after it was first introduced in Italy circa 1570. The Pitot tube is used in wind tunnel and airborne studies. It operates on the Bernoulli principle, Equation 4.11, which may be expressed as

$$p_s + (0.5\rho u^2) = p_0 \tag{5.1}$$

where $p_s$ is the hydrostatic pressure on a horizontal streamline of a moving fluid velocity, u, at a point (also termed the "static head" in the British Handbook of Meteorological Instrumentation, BMOAM, 1961); $p_o$ is the pressure at the stagnation point (i.e., u = 0; also known as the stagnation pressure, total pressure, or "total head" in BMOAM, 1961), and is a constant throughout the fluid. The remaining term is the dynamic pressure (named the "velocity head" in BMOAM, 1961). The windspeed is deduced from the difference between the dynamic wind pressure (suction force) and the hydrostatic wind pressure, $p_s$ (static force). The anemometers that fall under the aerodynamic and thermodynamic groups are discussed in detail by others (e.g., Middleton and Spilhaus, 1953; Wang, 1975; Fritschen and Gay, 1979; Brock, 1984).

Wind-measuring instruments are also classified according to the measurement platform (i.e. eulerian or lagrangian), the coordinate system represented by their platform, or the component of the windfield that they measure (i.e. direction or speed). They are also classified by the way they transform energy and frequency characteristics of the wind. Devices that transform the mechanical energy (aerodynamic), thermal energy (thermodynamic), or the electromagnetic spectral frequency shift (remote wind sensors) of the windflow past them are anemometers or windspeed-measuring devices. The eulerian platform involves velocity and gust measurements made by a wind device from a fixed platform with respect to the earth's surface. The lagrangian platform involves the continuous monitoring by the wind devices that follow the moving air current.

The eulerian systems are usually orthogonal cartesian coordinate systems, although polar and even cylindrical (curvilinear) coordinate systems are used. The wind device is considered to be in the center of the system at all times. Consequently, the wind components are identified as the wind passes through the system.

The lagrangian windfield measuring coordinate systems are determined primarily by ground or spacecraft observation of a moving subplatform or tracers. The moving subplatform is most commonly a balloon, although other "subplatforms" have been used (i.e., oil, flourescent particles, aerosols, visible plumes, and even moving cloud elements). There are many techniques available for ground detection of these moving objects—and, consequently, wind—the most common of which are radio-theodolites, remote sensing, and theodolites. Most of the radiosonde units commercially available today (e.g., Vaisala, 1995a) use telemetry, radio-theodolites, long range navigation (LORAN) (e.g., Colby, 1994), and global positioning (GPS) (Businger et al., 1996).

## *5.1. Wind Direction and Windspeed*

Wind direction is a parameter of direct interest in meteorological, environmental, and other applications, and is arguably one of the most important operational measurements. It has been measured for over two thousand years, apparently beginning as far back as Roman and Greek times, using a device known as a *wind vane*. The wind vane is essentially a body mounted so that it freely rotates about a vertical axis, and it indicates the direction of the resultant force on it due to the pressure of the wind that passes through its vertical axis. The classification of wind vanes is only based on their configuration and is detailed by Wierenga (1967). They simply come with or without propellers, and mostly without. The optimum dynamic performance of these wind direction indicators is ensured by:

(1) minimizing friction at their pivot point,
(2) exactly balancing the weights of the counterweight (sometimes called the mass) and the tail (sometimes called the fin),
(3) making sure the mast or support for the vane is vertical and true north is accurately determined for the location,

(4) maximizing the torque in relation to the vane's moment of inertia as the wind direction changes,
(5) minimizing the vane's moment of inertia, and
(6) using a slightly underdamped vane.

The minimization of friction at the vane's pivot point yields a better response for lower windspeeds. An exact balance of the mass and the fin ensure that the vane does not tilt in a certain direction while spinning. Minimizing the moment of inertia acts to minimize the aerodynamic effects on the vane during gusty periods. A slightly underdamped vane response minimizes the distance constant and maximizes the vane's response.

A wind vane will indicate true direction to within $\pm 1°$ (except at very low windspeeds) if the wind is steady (i.e., $u = \overline{u} + u'$ where $u' = 0$ and $\overline{u}$ is the mean wind). When the wind suddenly changes direction, the accuracy at any moment may be off by several degrees as the vane responds in a manner dependent upon the dynamical characteristics of its design (e.g., Brock, 1984). The dynamic characteristics of a vane include the starting threshold or delay distance, damping ratio, distance constant, frequency response, and phase shift. Some of these dynamic characteristics are addressed in section 5.2.2. Wang (1975), Fritschen and Gay (1979), and Brock (1984) provide additional details on wind vanes.

The hot-wire anemometer is the most common thermodynamic anemometer device, however the cup and gill anemometers are still the most popular. The hot-wire anemometer uses the cooling power of the wind to infer windspeed and is most often used in atmospheric field and laboratory research applications. A similar device known as the Kata anemometer is used in industry. Sonic, laser, and microwave anemometers are examples of remote wind sensors that use the electromagnetic spectral frequency shift due to the wind and have been referred to as frequency shift anemometers. The latest sonic windfield-measuring devices (e.g., Plate 5.1) measure both windfield and wind direction with virtually no moving parts. As a result, they are nearly maintenance free, less prone to icing errors, and more accurate.

The theory and operation of a cup anemometer and a common wind vane, hot-wire, and sonic, laser, and microwave anemometers are discussed in the following sections.

***5.1.1. Cup and Gill Anemometer and the Vane.*** The cup anemometer operates on the principle that a steady windspeed, u, causes a linear cup speed U (at cup centers) such that they are generally related by the power series expression

$$u = a + (bU) + (cU^2) + \dots + (nU^n) \tag{5.2a}$$

where a, b, c, ... , n are constants. The best design to date seems to occur when the coefficient a is constant and the coefficient of $U^2$ and higher powered terms of U are zero. The technology exists for finding optimal anemometer characteristics, but its transfer is slowly forthcoming. The National Aeronautics and Space Administration

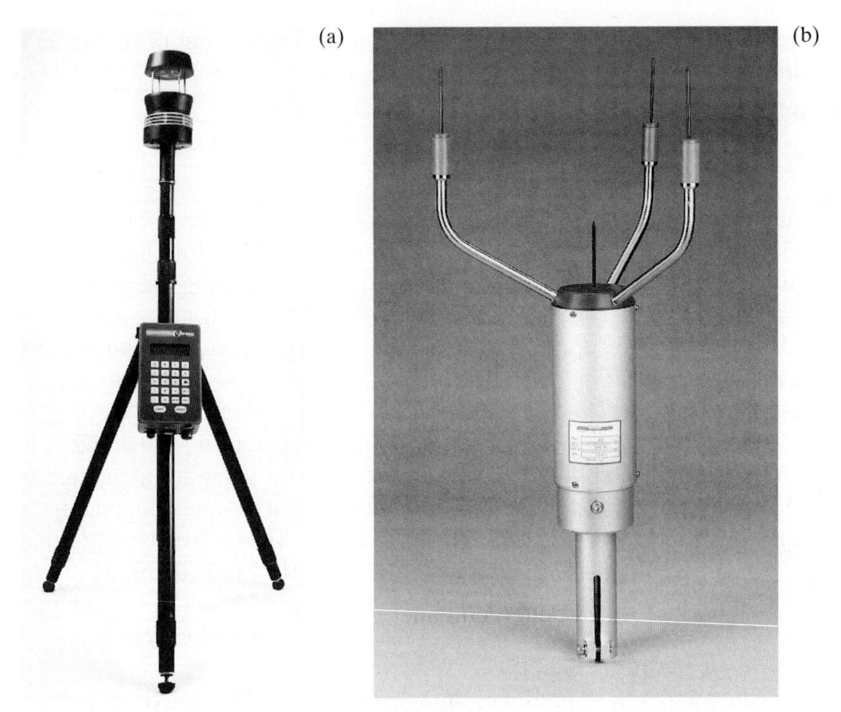

**Plate 5.1** *(a) The Climatronics Novel Sonic Anemometer. (b) A sonic anemometer from Handar (Lockyer, 1996). (Courtesy of D. Gilmore and D. Katz of Climatronics Corp. and K. Schlichting of Handar.)*

(NASA) Ames Research Center at Moffett Field, California is well on its way in perfecting the technology transfer process. Nonetheless, the results of controlled wind tunnel experiments and some field experiments indicate that the ideal anemometer configuration contains 3 evenly spaced conical-shaped, beaded cups whose center of mass is located at a distance from the center (axis) that is equal to the cup diameter. The conical cups yield a more linear response than cups of other shapes (i.e., hemispherical). The torque is maximized when the angle at which the wind strikes the anemometer (i.e., the angle of attack; also applies to a wind vane) on the concave surface is 45° (e.g. See Brock, 1984). The distance from the axis to the center of mass of the cup is known as the arm length. Beaded cups minimize the effects of turbulence (a second order or higher term in Equation 5.2a). The smaller and lighter the cup, the lower the moment of inertia and friction of the device (second order or higher terms in Equation 5.2a). The ratio of the cup diameter to 2 times the arm length is the anemometer factor (Brazier, 1914). An anemometer factor of 0.5 contributes to the linearization of the relationship between u and U, namely (Equation 5.2a) becomes

$$u = a + (bU), \tag{5.2b}$$

where a becomes the threshold value of the anemometer (i.e., the maximum value of U for which u is zero), and b is the slope of the regression line, or $(\partial u\ \partial U^{-1})$.

The behavior of cup anemometers in gusty winds has been theoretically analyzed by Schrenk and others particularly in wind tunnels since 1892 (e.g. BMOAM, 1961; Brock, 1984). Observations show that as the windspeed increases, the cup wheel of the anemometer accelerates more quickly than it deaccelerates when the windspeed decreases. The latter yields a mean speed recorded by the anemometer in a gusty or variable wind that is greater than the "true" mean wind. Shrenk related the over-estimation of the cup anemometer with a nondimensional parameter denoted as K, given by

$$K = 0.55\rho R^2 r^2 T u I^{-1} \tag{5.3}$$

where T is the period of the windspeed variation (assuming the oscillation is sinusoidal); u is the mean windspeed; $\rho$ is the air density; R is the radius of the wheel, i.e., circle described by the center of the cups; r is the radius of the cups, and I is the moment of inertia of the rotating parts.

The gill anemometer is essentially a freely moving propeller attached to a stick whose rotation is translated into a torque that is then used to obtain the windspeed of the air moving past the propeller. Gill anemometers can be directional, and are discussed in detail elsewhere (e.g., BMOAM, 1961; Wang, 1975; Fritschen and Gay, 1979, Brock, 1984). The direction may, for example, be measured with the gill anemometer by allowing it to rotate freely in the azimuthal direction. The gill-type U,V,W anemometer (Fig. 5.3) simultaneously measures the three orthogonal vectors of the windspeed, namely, east–west (U), north–south (V), and vertical (W), with three independent sensors (usually helicoid propellers). The propellers are mounted at right angles to each other on a mast. They rotate $\approx$3.15 revolutions $m^{-1}$ ($\approx$0.96 revolutions $ft^{-1}$) for all windspeeds $>\approx$1.2 m $s^{-1}$ ($>\approx$2.7 mi $h^{-1}$), with a threshold value of $\approx$0.36 m $s^{-1}$ ($\approx$0.8 mi $h^{-1}$) and a maximum of $\approx$22.4 m $s^{-1}$ ($\approx$50 mi $h^{-1}$), according to the manufacturer's specifications.

***5.1.2. Hot-Wire Anemometer.*** The modern hot-wire anemometer is usually made with one of two electronic bridge configurations, namely constant current or constant temperature. The constant current configuration measures the windspeed as follows. A given increase in windspeed enhances the heat transfer to the environment, which in turn decreases the temperature and resistance of the wire. The resistance decrease results in a voltage imbalance of the bridge circuit that, when amplified within a non-linear amplifier, is an indication of windspeed. The constant temperature or constant resistance configuration works in a similar fashion. A given increase in the windspeed past the sensor results in cooling and decreases the resistance. The decrease in resistance causes a decrease in the bridge output which, when amplified, causes an increase in current to the sensor. This increased current then returns the resistance to its original value, $R_{wo}$. The overheat ratio is typically 0.5 and is defined in terms of a ratio of the original or preset resistance of the sensor and the resistance at ambient temperature. If the

**Fig. 5.3** *Photo of a Gill-type U,V,W anemometer. (Courtesy of Qualimetrics, Inc.)*

amplifier has sufficient gain, it will tend to keep its inputs very close to balance. The current that maintains the constant resistance is used as the system output, since it is a function of windspeed, as will be seen below (e.g., Equation 5.9). This configuration:

(1) prevents sensor burnout that may occur with a constant current system during a lull in wind
(2) allows for linearization
(3) can be temperature compensated, and
(4) yields a direct DC output.

The hot-wire or hot-film anemometer (Fig. 5.4) operates on the principle that the rate of heat loss from a body is related to the rate of airflow past it. The rate of heat loss from the hot wire sensor depends on the temperature, geometrical shape and dimensions of the sensor, as well as the velocity, temperature, pressure, density, and thermodynamic properties of the air. Given that only the velocity of the fluid is

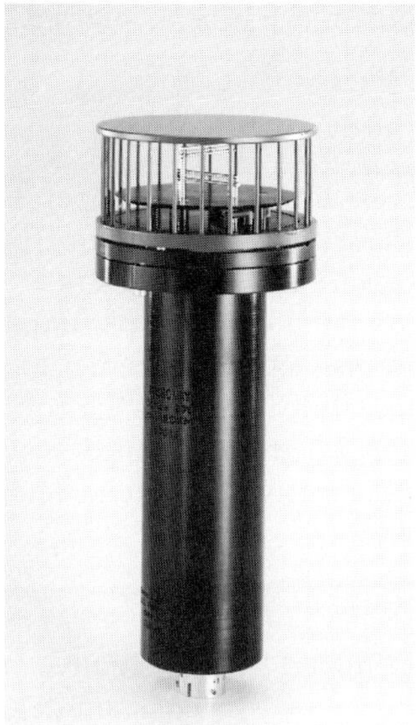

**Fig. 5.4** *Photos of three examples of hot-wire anemometers. The Model 500 (far right) is a hot-film anemometer. (Courtesy of Dave Gulati, and Cossonay Meteorology Systems, a product line of Fidelity Technologies Corporation)*

allowed to vary, then the heat loss is a measure of the velocity change. A simple heat budget describing all of the losses of heat from the sensor can account for the behavior of the hot-wire or hot-film sensors. The total rate of heat transfer, $Q'$, from a hot-wire or hot-film with resistance, R, when a current, I, passes through it may be expressed in terms of the forced convective heat loss, $Q'_c$, the conductive heat loss from the wire support, $Q'_s$, the radiative heat loss, $Q'_r$, and the rate of change of heat content, $Q'_m$ at equilibrium as

$$I^2R = Q' = Q'_c + Q'_s + Q'_r + Q'_m. \tag{5.4}$$

The $Q'_c$ may be expressed in terms of the windspeed, u, and, assuming a cylindrical geometry, as

$$Q'_c = \pi l K Nu \Delta T, \tag{5.5}$$

where l is the sensor length, K is the thermal conductivity of the air, $\Delta T$ is the sensor temperature minus air temperature also known as the overheat, and Nu is the Nusselt number. The rate of heat transfer in the free convection regime and at low velocities reduces to

$$Q' = Nu K (T_s - T_a), \tag{5.6}$$

where $T_a$ is the air temperature $T_s$ is the heated element temperature, and Nu is a function of the Reynolds number, Re, Prandtl number, Pr, and the Grashof number, Gr. The laminar flow cases in a free convection regime employ a Nusselt number expression that is applicable for cylindrical, spherical, and vertical plate flow geometries in the form

$$Nu = 0.54(Gr\,Pr)^{0.25}, \quad \text{that becomes} \quad Nu = 0.66\,Re^{0.5}\,Pr^{0.333}$$

for flat surfaces in the environmental temperature range and laminar flow. Additional details of the theory may be found in Hinze (1959), for example.

The operation of hot-wire anemometers (see Wang, 1975; Fritchen and Gay, 1979 and Brock, 1984 for details) entails a platinum wire (King, 1914, used 0.003" diameter) that is heated by an electrical current, i, and held perpendicular to the airflow of speed, U, and temperature $T_a$. When the equilibrium temperature, $T_e$, of the wire is reached, equilibrium electric resistance, $R_e$, is obtained. Consequently, the equilibrium relationship between wire element temperature, air temperature, and air velocity may be given by:

$$T_e - T_a = R_e i^2 [k_1 + (k_2 u^{0.5})]^{-1}, \tag{5.7}$$

where $k_1$ is a constant determined by the radiative and convective heat losses from the heated wire when $u = 0$, and $k_2$ is a constant determined by the diameter of the wire and the physical properties of the ambient air. $k_1$ and $k_2$ may be determined empirically. The total heat loss from the sensor of diameter, D, at equilibrium in terms of power, P, consumed, is

$$Q' \pi D l = P = i^2 R_e, \tag{5.8}$$

where l is length of wire, and $Q'$ is the total heat loss rate. The equation for the heat transfer in air may be expressed in terms of the Nu and Re of the flow past the wire. The Nusselt number may be given by

$$Nu = [0.24 + (0.56\,Re^{0.45})][(T_s + T_a)(2T_a)^{-1}]^{0.17} \tag{5.9}$$

over the range $0.02 < Re < 44$, where $T_s$ is the actual (not necessarily equilibrium) temperature of the sensor.

Individual sensors need to be calibrated when their length is short enough that there is heat transferred to the mounting stubs. The calibration factor for a hot-wire anemometer at operating temperature resistance, R, and a resistance at ambient temperature of $R_a$, takes the form

$$(i^2 R)(R - R_a)^{-1} = A + (B u^n) \tag{5.10}$$

where A, B and n are constants determined experimentally. When the platinum wire is heated to a very high temperature (i.e., several hundred degrees celsius), then $T_e - T_a \approx T_e$. The placement of the wire in a wheatstone bridge could be used to keep $T_e$ and $R_e$ constant. This would simplify (Equations 5.6 to 5.9) to

$$i^2 = i_0^2 + (K_e u^{0.5}), \tag{5.11}$$

where $i_o$ is the electric current measured in still air and is a constant, and $K_e$ is another constant. This equation represents King's law and shows that i is a function of windspeed. Furthermore, if current, i, is kept constant, then the resistance of the wire, or the electric potential drop across the wire, will be a function of windspeed at the equilibrium point.

The form of output is voltage (voltmeter or potentiometric recorder), or current (current meter, voltmeter, or recorder). The time constant may be as short as 0.1 s and the range as wide as 0.1 to 22 mph. The lag coefficient, $\lambda$, is also very small and may be obtained from

$$\lambda = 4.2\,\rho_{wi} A^2 C_{wi} (T_e - T_s)(R_{wo} i^2)^{-1}, \tag{5.12}$$

where $\rho_{wi}$ is the density of the wire, A is the cross-sectional area of the wire, $C_{wi}$ is the specific heat of the wire, and $R_{wo}$ is the specific resistance of the wire at temperature $T_s$ (after BMOAM, 1961). The most rapid response requires the use of a very fine wire and as small a temperature excess as possible.

The hot-wire anemometer is significantly more accurate and quicker to respond to wind fluctuations than the aerodynamic anemometer. It is also more sensitive for low windspeeds if a very fine wire is used (e.g., 1 cm long and 2–5 $\mu$m in diameter). The sensing element of a hot-wire anemometer is usually a 0.013 to 0.13 mm platinum wire with an operating temperature between 200 ° and 500 °C. A Tungsten wire of 5 $\mu$m diameter and 1.5 mm length is also used. The hot-wire anemometer operates best when its wire has a dull red glow (Brock, 1984).

A hot-film sensor may be used instead of a wire to measure windspeed. The hot-film sensor consists of a thin layer of conductor (such as platinum) deposited on a quartz rod of diameter between 25 to 250 $\mu$m and length about 1.5 mm. The hot-wire or hot-film anemometer is used to measure windspeeds from a few mm $s^{-1}$ to supersonic. It is lightweight, has a quick response to changes in fluid velocity (generally $\leq$1 sec for 63% of signal change) at frequencies up to thousands of hertz, and their accuracy is $\pm$2 cm $s^{-1}$ ($\approx$0.04 mi $h^{-1}$; or 8% of mid-scale for typical Japanese and Finlandese versions). The response frequencies are commonly in the hundreds of hertz for hot-film anemometers.

Unfortunately, winds cannot be measured using a hot-wire anemometer during wet weather conditions (i.e., rain, cloud, or fog), and the hot-wire anemometer is more fragile than a conventional anemometer. Another disadvantage is that its calibration is variable (i.e., primarily due to wire contamination). The hot-films are less susceptible to contamination problems, have a lower conduction to the support, and a better frequency response (approaches 1 MHz). Nonetheless, the hot-wire or hot-film anemometer is best suited for wind tunnel research.

### *5.1.3. Remote Wind Sensors*

*Sonic Anemometer.* Sonic anemometers have been commercially available for a few decades or so. They are fast becoming reasonably affordable due to improvements in electronics and their increased commercial availability (i.e., Applied Technologies in Boulder, Colorado; Climatronics in Bohemia, New York; COSSONAY Meteorology Systems SA in Penthalez, Switzerland via Fidelity Technologies Corp. in Reading, Pennsylvania; Handar, Inc. in Sunnyvale, California; and Mesa Systems in Framingham, Massachusetts, are among those that now sell these instruments).

The sonic anemometer passively measures the windspeed (see Chapter 2) using a sound pulse or continuous wave configuration.

The principle of operation of sonic anemometers simply entails the interference in the frequency of sound pulses sent across a short path length due to wind. Briefly, the time difference, $\Delta t$, between the initial transmission of sound waves across the air stream and their reception (i.e. return) is a direct function of the mean air speed along the path, namely

$$\Delta t = 2 l u \cos \Theta (c^2 - u^2)^{-1}, \tag{5.13}$$

where l is the acoustic path length, and $\Theta$ is the angle of the wind (of speed u) with respect to the sound wave (of speed c). The mean of the return pulses is proportional to the air temperature. The continuous wave sonic anemometers operate in a similar manner.

Sonic anemometers are inertia free, can be used to measure over very short or long path lengths, and may measure three orthogonal wind components. For example, Kaimal and Gaynor (1983) used sonic anemometers to obtain vertical velocity. The resolution of sonic anemometers is slightly better than 0.03 m $s^{-1}$ over an operating range of 0 to 30 m $s^{-1}$, and their time constants are on the order of 0.01 s. They have a quicker time constant than the hot-wire anemometers, and one less component that might need replacement.

The path length over which the speed is measured may be designed according to ones fancy, meaning that 30 m $s^{-1}$ is not necessarily the maximum windspeed that can be measured by a sonic anemometer. Actually, the path length is usually ≤15 m (≈30 ft) and can be as long as 1500 m (≈5,000 ft) without overly significant interference (i.e., distortion errors) from strong turbulence, temperature and wind gradients, heavy rain, etc. The speed of sound has a slight dependence on temperature. For example, the speed of sound varies from ≈360 m $s^{-1}$ at +50 °C to ≈300 m $s^{-1}$ at −50 °C, and it should be noted that this change is on the order of the range of most sonic anemometers. The speed of sound is also very slightly dependent on water vapor pressure and air pressure. The data collection rate for a typical sonic anemometer is generally 200 measurements per second (Brock, 1984), which, as you might imagine, creates a very large amount of data in a continuous mode operation. Please note that the 200 measurements per second, for example, are typically averaged to 10 Hz or 20 Hz values, depending on the application. A slow response mode for a typical sonic anemometer is 1 measurement $s^{-1}$.

Consider, for example, two sonic anemometers, one from Climatronics and another from Handar, Inc.

The Climatronics novel sonic anemometer (Plate 5.1) described by Robertson and Katz, 1995) comes with two degrees of accuracy and sensitivity depending on the application. Both are designed to replace the generic cup and anemometer vane systems (Figs. 5.1 and 5.2). One version has a range of 0 to 50 or 60 m $s^{-1}$ with ±0.5 m $s^{-1}$ or ±5% accuracy, a path length of 10 cm, and the other has a more rugged sensor with a range of 0 m $s^{-1}$ to 65 m $s^{-1}$ ±1 m $s^{-1}$ or ±10% accuracy. Here, rugged means that the instrument is immersible in seawater for short periods and can be hosed down for decontamination if it is exposed to dangerous chemicals. Either version can be easily integrated into complete automated weather stations.

The Climatronics novel sonic anemometer, or another manufacturer's sonic anemometer is (a) as accurate, and more rugged than cup and vane and propeller anemometers, (b) kept ice free with a reasonable power level, (c) more stable and consumes less power ($<0.5$ W) than the hot-wire anemometer, and (d) as costly as the cup and gill anemometers (Robertson and Katz, 1995).

The Handar ultrasonic wind sensor (Plate 5.1) measures windspeed at an accuracy of $\pm 0.1$ m s$^{-1}$ for "true" windspeeds below 4.5 m s$^{-1}$, and 3% of the reading for true windspeeds $\geqslant 4.5$ m s$^{-1}$(Lockyer, 1996). Its wind direction accuracy is $\pm 2°$, and it costs as much as their high quality mechanical wind sensor (REMS•Talk, 1995). This sonic anemometer does not have a plate at the top as do the others.

*Laser Anemometer.* The laser anemometer consists of a laser beam that shines on moving light scattering particles. A receiver unit detects the backscattered portion of the initial beam and the electronic circuitry measures the resulting frequency shift (Doppler effect) which is relatable to the windfield. These anemometers are presently becoming popular and may be used on ground, airborne or ultimately satellite platforms. For example, the $CO_2$ Doppler wind lidar has been used on the ground and in the air since 1968. Its successful use led to the development of the Laser Atmospheric Wind Sounder (LAWS) instrument or a modern type of Doppler lidar in 1985 for NASA, Earth Observing System (EOS), that may ultimately be deployed on a satellite platform.

The LAWS (Fig. 5.5) operates on principles similar to Doppler radars. Simply, photons are sent into the atmosphere from the LAWS's transmitter or laser. These photons are scattered back to its receiver as they travel through the atmosphere, and the return signals are spectrally analyzed to recover Doppler shifts due to the motions of the targets (Baker et al., 1995). The targets are much smaller than those for radars, consequently, the photons from the lidar have shorter wavelengths (namely, $\approx 9$ $\mu$m for the LAWS, and $\approx 11$ cm for a typical weather Doppler radar). The LAWS targets are cloud particles or suspended atmospheric aerosols that move nearly as fast as the wind. The small targets require that the operational signal-to-noise ratio be at least $10^{-11}$ m$^{-1}$ sr$^{-1}$, in order to retrieve the windspeed, especially for space-based lidar systems (Baker et al., 1995). Note too that the spaceborne platform has a ground speed of at least 7 m s$^{-1}$ (Baker et al., 1995) that must be taken into account. Baker et al. (1995) also discuss how to optimize the parameters of the LAWS to minimize the errors with respect to the particular requirements. They show that the total variance $Ó_0^2$ of the error in the retrieved LAWS windspeed can be obtained from

$$Ó_0^2 = (Ó_n^2 + Ó_s^2)(pN)^{-1} \tag{5.14}$$

where $Ó_n^2$ is the variance in the line of sight measurement due to the lidar system noise, atmospheric turbulence within the sample volume, stability of the wind estimation algorithm, etc. and is generally $\approx 0.5$ m s$^{-1}$. $Ó_s^2$ is the variance of the true wind component on scales between 1 and 200 km (assuming a $10^4$ km volume with a 1 km

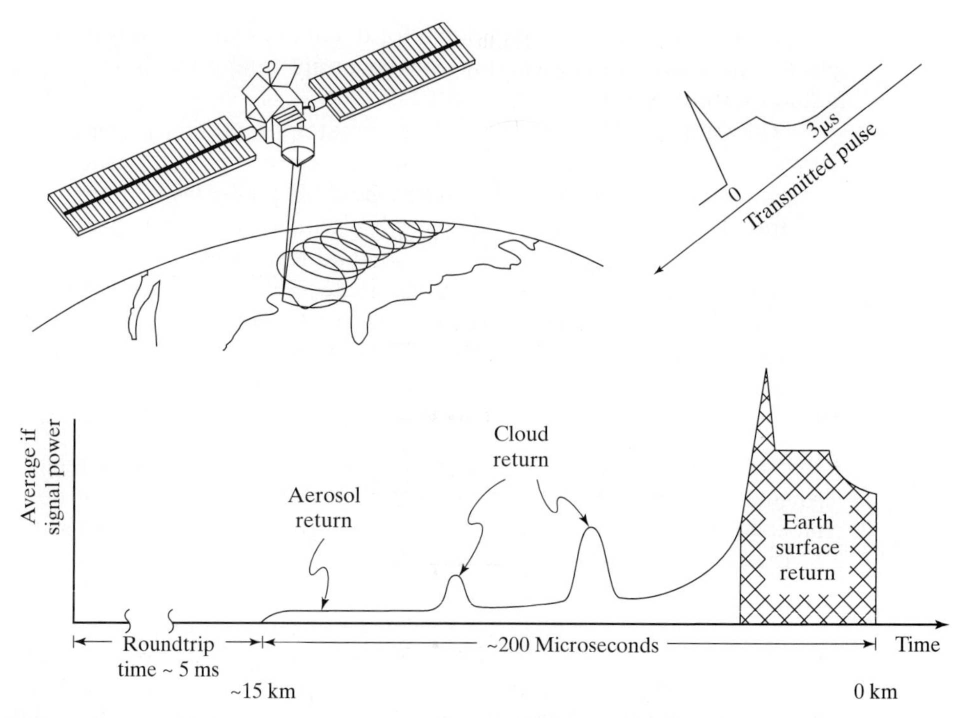


**Fig. 5.5** *A Laser Atmospheric Wind Sounder, LAWS. (Adapted from Baker et al., 1995 with permission from American Meteorological Society)*

depth) and is generally ≈3 m s$^{-1}$. N is the number of attempted line of sight observations from the total sample of M observations that lie within a specified error tolerance, f; p is the fraction of N attempts that resulted in a "good" line of sight estimate. Baker et al. (1995) state that $Ó_n^2$ and p are partial functions of the laser pulse energy, while $Ó_s^2$ is independent of laser energy and may be reduced only by increasing the product of p and N.

Laser anemometers provide very accurate wind measurements (e.g. Table 5.1), and they may also yield information on cloudtop height, cloud thickness, closure on atmospheric-ocean heat transport, planetary scale dynamics and climate, aerosols, weather prediction, and dynamics of weather systems, among others.

*Microwave Anemometer.* The special sensor microwave imager (SSM/I) aboard a Defense Meteorological Satellite Program (DMSP) satellite has been said to determine wind vectors near the surface (e.g., Wentz, 1992; Atlas et al., 1996), with very promising preliminary results. In fact, these data are currently being used in a variety of atmospheric and oceanic applications (Atlas et al., 1996), such as the retrieval of surface temperature, atmospheric vapor content, and clouds and precipitation characteristics.

*Wind Profilers.* Radar wind profilers (Fig. 5.6) are also another possible way to determine the windfield within the troposphere. They yield continuous, unattended

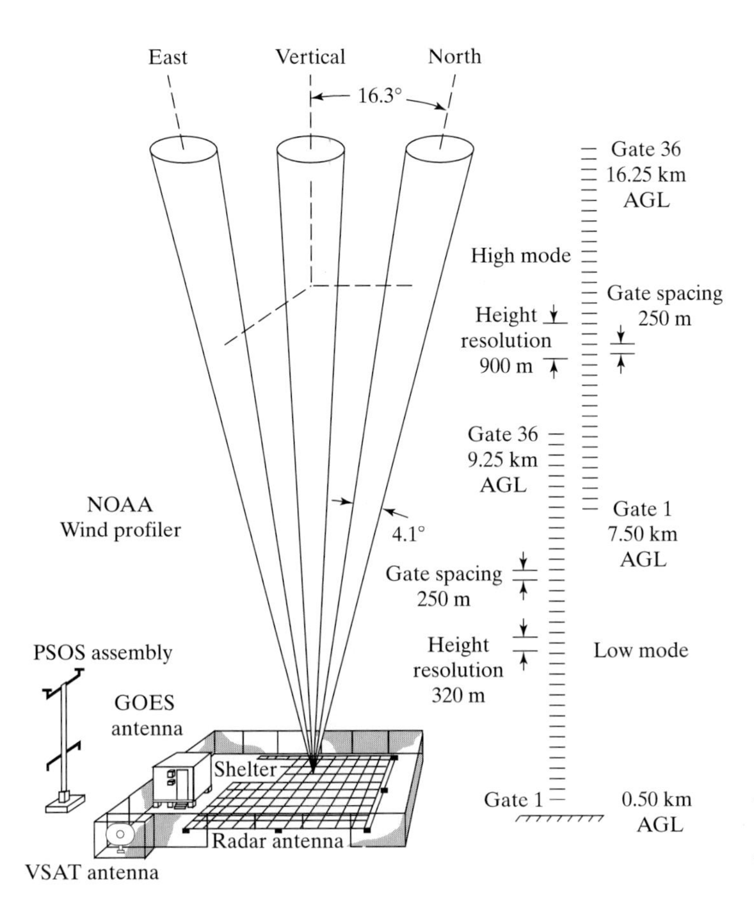


**Fig. 5.6** *A radar wind profiler. (Adapted from Ralph et al., 1995 with permission from American Meteorological Society)*

vertical profiles of the windfield that may be averaged over a desired time interval. Radar wind profilers simply detect backscattered signals from turbulence-induced refractive index variations with a scale of half the radar wavelength. As the refractive index irregularities flow with the mean wind, their translational velocity provides a direct measure of the mean wind vector. The mean data values are fit by a series of sine and cosine terms to determine the amplitudes and phases of the harmonic terms (known as a harmonic analysis). The harmonic analysis is done on the east–west and north–south wind components in order to isolate the wind oscillation. So, for n data points in a period T, the highest frequency waves that can be resolved by discrete Fourier transforms are n/2 waves. Thus if n = 24 (i.e., 1 data point per day), then 12 harmonic components with frequencies between 1 and 12 waves per day can exactly fit the 24 data points, such that

$$
\begin{aligned}
u(t) &= u_0 + \sum_{1}^{n/2} [A_{un} \sin((nt) + \eta_{un})], \\
v(t) &= v_0 + \sum_{1}^{n/2} [A_{vn} \sin((nt) + \eta_{vn})],
\end{aligned}
\tag{5.15}
$$

where $u_o$ and $v_o$ are the mean wind components over the period T (in this case 24 hours), $A_{un}$ and $A_{vn}$ are the amplitudes, and $\eta_{un}$ and $\eta_{vn}$ are the initial phases of the nth harmonic terms for u and v. The components in Equation 5.15 are added as vectors to obtain windfield quantities and a power spectrum since velocity is related to power (Fig. 5.7). The vector averaging that is used to preprocess the profiler data provides an effective filter that allows harmonic components to pass but filters out all other oscillations, including synoptic wind oscillations. The longer the data record, the better the filter.

The radar wind profilers in the United States operate on three frequencies: 915, 404, and 50 MHz. Whiteman and Bian (1996) provide additional details. The 50 MHz profilers can obtain a range of 20 km under ideal conditions, whereas the 404 MHz profilers have a maximum range of 16 km under ideal conditions (Doviak and Zrnic, 1993). Whiteman and Bian (1996) found the typical range of the profilers to be closer to half of the aforementioned maximum values, and that the lower frequency profilers have a poorer range resolution than that of the

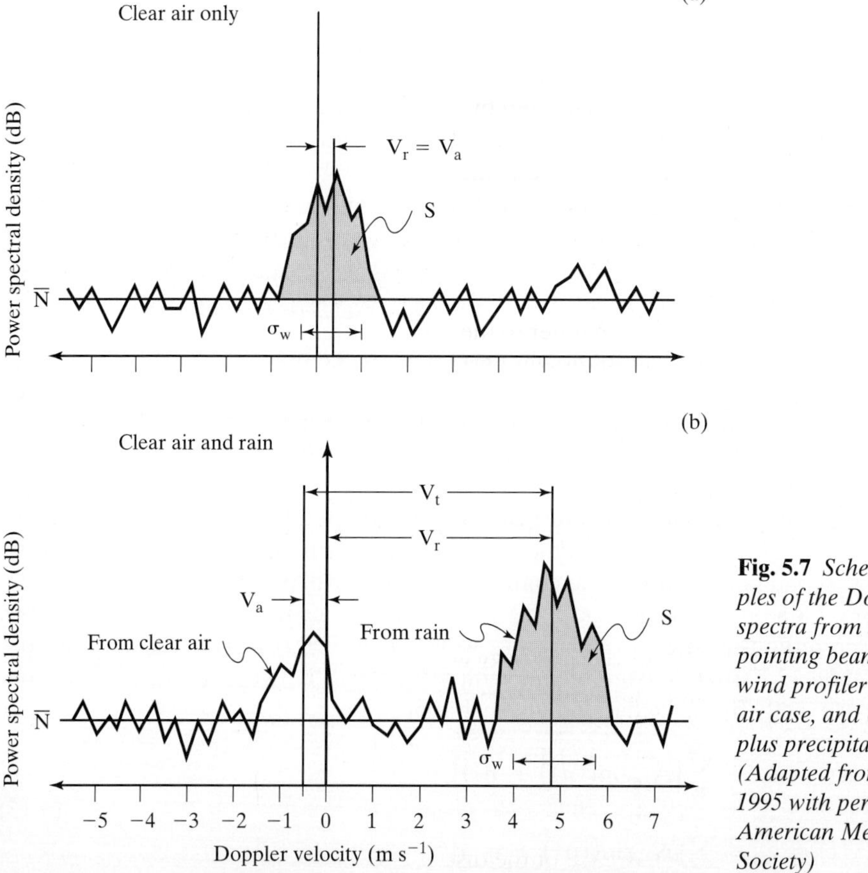


**Fig. 5.7** *Schematic examples of the Doppler power spectra from the vertically pointing beam of a radar wind profiler for (a) a clear air case, and (b) a clear air plus precipitation case. (Adapted from Ralph et al., 1995 with permission from American Meteorological Society)*

larger frequency profilers. For example, their 404 MHz profiler data have a resolution of 250 m altitude, whereas their 915 MHz profiler data have a resolution of 103 m altitude intervals. Whiteman and Bian (1996) also found that the profiler amplitude errors, although preliminary, were generally between 0.1 and 0.2 m $s^{-1}$, not including the ranges near the ground (i.e., ≤500 m above ground level, AGL, for the 404 MHz data and ≤177 m AGL for the 915 MHz data) and at higher altitudes. Furger et al. (1995) reported absolute accuracies of 915 MHz radar profilers to be ≈1 m $s^{-1}$ when the data was averaged over a climatological mean day before performing a harmonic analysis. Neff (1990), Doviak and Zrnic (1993), Wurman (1994), Cohn and Chilson (1995), Furger et al. (1995), and Whiteman and Bian (1996) provide additional details.

There is a related device known as the Doppler sodar which has been used in air pollution studies over the past 20 years according to Crescenti (1997) who details their theory of operation, advantages and disadvantages. Crescenti (1997), however, states that while these devices can provide reliable windfield profiles, he cautions against their use, stating that they are not ready for practical use.

## 5.2. *Response Times*

***5.2.1. Windspeed.*** The ability of the wind-measuring device to respond to changes in the windspeed is quantified by the distance constant and the threshold of the wind-measuring device, or their equivalents. The distance constant, L, sometimes referred to as the response distance is the length of an air column required to pass an anemometer with speed, u, in order to allow the anemometer to respond to 63% of the step change in speed from the initial to final condition. Mazzarella (1972) showed the distance constant for cup anemometers to range between 2.5 and 26.2 ft. The time required to reach the distance constant is called the time constant, $\tau$, and is determined in an analogous manner to the time constant of a thermometer. The time constant of a cup anemometer, assuming a first order system, may be mathematically expressed as

$$\tau( dU\, dt^{-1}) + U = f(t) \tag{5.16}$$

where U is the linear speed of the rotating cups, t is the time in seconds and f(t) is the instrument constant. f(t) = 0 when t ≤ 0, and f(t) = k when t > 0. The general solution of (Equation 5.16) is

$$U = k[1 - e^{-(t\tau^{-1})}]. \tag{5.17a}$$

The time constant ($\tau$) increases inversely with windspeed (u) if the frictional force of the anemometer is negligible and the second order Reynolds number effects are ignored. Given that the windspeed changes from $u_o$ to $u_o + \partial u$, then using $u_o$ as the equilibrium value, the linear speed of the rotating cup U will be

$$U = \partial u[1 - e^{-(t\tau^{-1})}] = \partial u\,[1 - e^{-(xL^{-1})}], \tag{5.17b}$$

where x is the length or distance of the displacement of the wind during time t.

**Table 5.1** *A comparison of several types of anemometers*

| General Type | Time Constant | Required Electronics[a] | Accuracy ($m\ s^{-1}/°$) | Relative Cost[b] | Threshold ($m\ s^{-1}$) |
|---|---|---|---|---|---|
| Rotating | seconds to minutes | none | ±1/— | L | ≈0.2–2.2 |
| Sonic | ⩽tenths of seconds | digital | ±0.02–1/2–8 | L to H | 0.02–1 |
| Wire | hundreds–thousands Hz | digital | ±0.02–1/— | vH | 0.02–1 |
| Lidar(LAWS) | ⩽tenths of seconds | digital | ±1–5/— | NA | 1–5 |
| Microwave | ⩽tenths of seconds to hours | digital | ±2/20 | NA | 2 |

[a] Digital here denotes complex digital electronics.

[b] Here L = low, H = high, vH = very high cost. NA = not applicable at this time.

The speed at which an anemometer starts to operate defines its threshold. For example, the threshold of a cup or propeller anemometer is the speed of air required to start the cups or propeller rotating, respectively. For example, a cup anemometer having an arm length of 7.5 cm and cup diameter of 5 cm has a threshold value of ≈0.5 mi. $h^{-1}$ (≈0.2 m $s^{-1}$). A cup anemometer having an arm length of ≈39 cm and cup diameter of ⩾10 cm has a threshold value of ≈5 mi. $h^{-1}$ (≈2 m $s^{-1}$). The latter anemometer example is approximately the size of the large anemometer used around 1970 by the United States National Weather Service. Table 5.1 shows the starting speed or threshold for some anemometers. The threshold value is useful when trying to determine whether an anemometer is better suited for measuring low or high windspeeds. A low threshold value (i.e., 0.5 mi. $h^{-1}$) indicates that such an anemometer is suitable for low windspeed measurements, while a high threshold (i.e. 5.0 mi. $h^{-1}$) indicates suitability for higher windspeed measurements. Friction is the only force to decelerate a spinning anemometer. Remember that the actual instrument selection depends on a number of additional factors, especially your operational requirement. For instance, rotating anemometers, particularly cup anemometers, are unreliable in variable winds because of their different responses for wind increase or decrease, whereas a sonic anemometer would be more reliable. Table 5.1 also clearly compares the response speed, electronical requirements, accuracy, and relative cost of some anemometers.

***5.2.2 Wind Direction.*** A wind vane is exposed to nearly continuous unbalanced wind forces resulting from unsteady wind flow that cause the vane to overshoot and undershoot the true wind direction in an oscillatory manner. As the vane approaches the true indication of the direction from which the wind is blowing, the oscillations will decrease. The reduction of oscillating amplitudes with time is known as "damping." The process is complicated by relationships between the forced and natural oscillations of the vane itself, and the fluctuations of the wind involving the inertial force and acceleration of the vane. Wind tunnel experiments indicate that the time required for the vane to indicate true wind direction is a function of windspeed, deflection angle, and the structure of the vane. The position of the shaft is typically monitored by a potentiometer in which electrical resistance varies in proportion to the wind direction.

Essentially all wind vanes are slightly underdamped meaning that it requires more time to reach its equilibrium position or true wind direction than it would if it were critically damped. A critically damped vane responds exponentially to a change in direction, and does not make for a sensitive wind vane, since it exhibits less amplitude, less oscillation and no overshooting. A distance constant can be defined for a critically damped vane, and the time required to respond to the step change can be quite large (Fritschen and Gay, 1979). Given the underdamped case, the oscillations of the vane form a decay curve of simple harmonic motion with one or two overshoots. An overdamped vane responds in an aperiodic manner and has an infinite period in its oscillations. The derivation of the equation of vane motion is left as an exercise for the reader, and Appendix C provides a reasonable beginning. The equation of the response, $\theta$, of this wind vane to indicate the "true" horizontal position after it is released from its initial position, $\theta_o$, may be represented by

$$\theta = \theta_0 e^{-(t/\lambda)} \sin\left[\phi + (2\pi t T'^{-1})\right], \tag{5.18}$$

where $\phi$ is the phase shift, $T'$ is the period of oscillation, and t is the time assuming small deviations of the vane position relative to the "true" wind direction (BMOAM, 1961). $\lambda$ and $T'$ depend on the windspeed and the dimensions and moment of inertia of the vane. $\lambda$ is essentially analogous to the time constant of a thermometer. In general, the total amplitude of the oscillations is reduced to 10% of its initial value in 2.3 $\lambda$ seconds.

Figure 5.8a illustrates the approach of an underdamped vane to its final true value ($\theta_f$) with respect to time. The amplitude of the overshoot of the vane is indicated by the X's, and is used to define a quantity termed the "damping ratio", $\xi$. The general form of the defining expression for the damping ratio is

$$2\pi\xi(1 - \xi^2)^{-0.5} = \ln(X_1 X_2^{-1}). \tag{5.19}$$

When the vane is critically damped, $\xi = 1$, and there is no overshoot. When it is overdamped, $\xi > 1$ and the oscillations are aperiodic. The underdamped condition yields $\xi < 1$ and there are usually a few overshoots that last for several seconds. The greater the damping ratio, the smaller the second overshoot amplitude. Vanes with a damping ratio of $\geqslant 0.4$ do not have a second overshoot, and $X_1$ and $X_2$ are obtained as shown in Fig. 5.8b. In this latter case, Equation 5.19 may be taken over $\pi$ instead of $2\pi$, and it reduces to

$$\xi = (1 + [\pi^2[\ln(X_1 X_2^{-1})]^{-2}]\}^{-0.5}. \tag{5.20}$$

Vanes with a damping ratio <0.4, may have three or four overshoots. The damping ratio is related to the natural and damped natural wavelengths of the vane's response to windfield by:

$$\xi = [1 - (\lambda_n \lambda_d^{-1})^2]^{0.5} \quad \text{and} \quad \lambda_d = d'[6.0 - (2.4\xi)](1 - \xi^2)^{-0.5}, \tag{5.21}$$

where $\lambda_n$ is the natural wavelength (the wavelength of a sinusoidal wave), $\lambda_d$ is the damped natural wavelength, and d' is the delay distance (see Fig. 5.9). The delay distance is the 50% recovery of a directional change, and is determined by the product of the tunnel windspeed, u, and the time for the wind vane to move from point of release to 50% of the distance where it first crosses the equilibrium line (or true

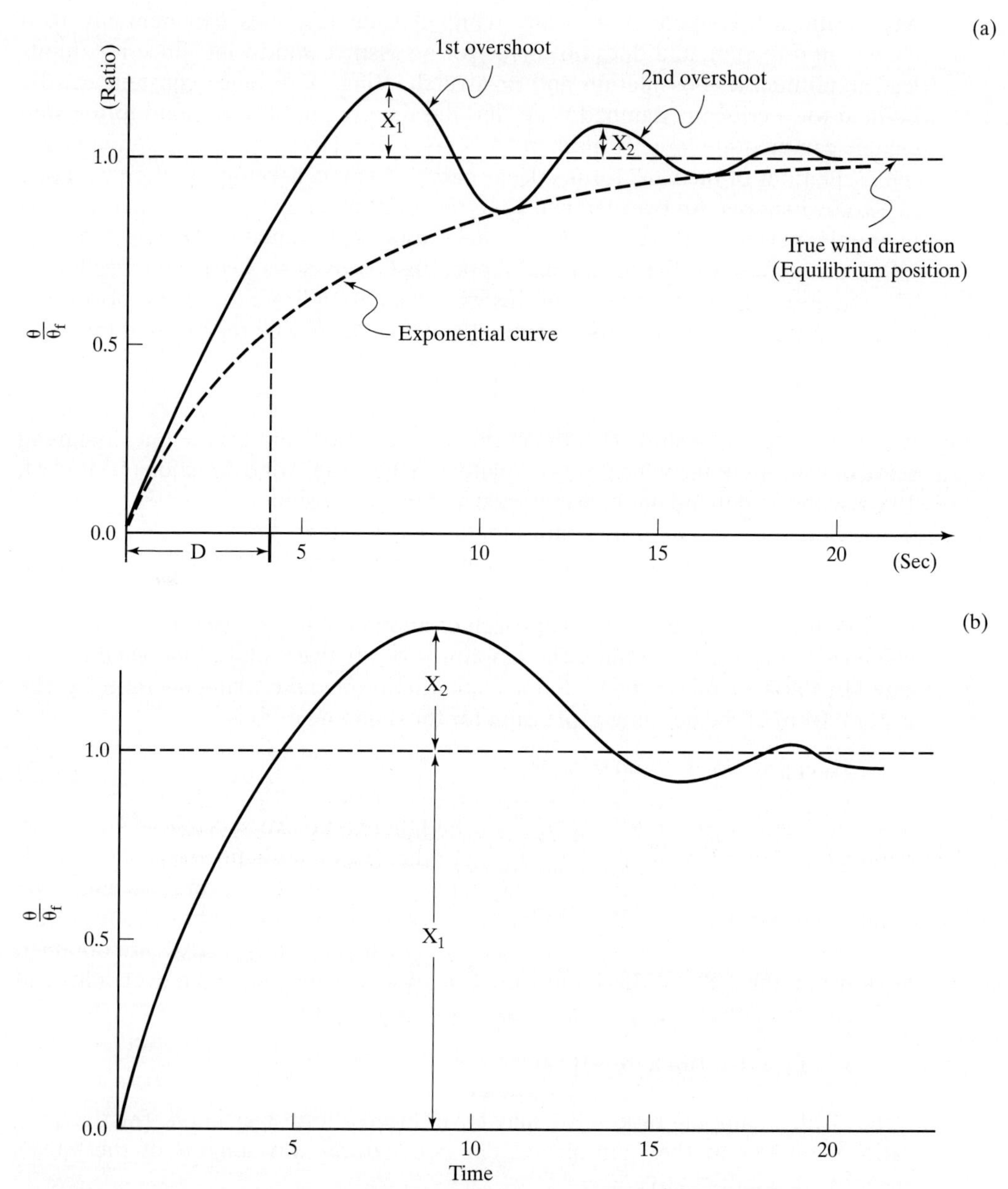


**Fig. 5.8** *(a) A vane response to wind flow that is underdamped. (b) Same as (a) except without the second overshoot. (After Wang, 1975, and Fritschen and Gay, 1979.)*

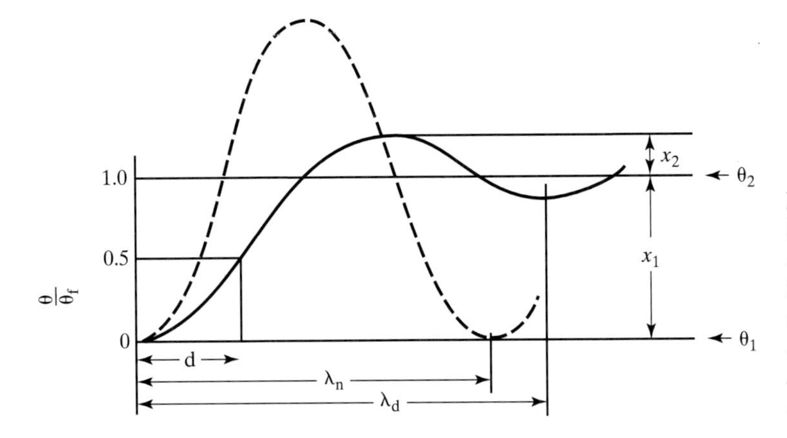


**Fig. 5.9** *The relationship between the natural and damped natural wavelengths, and the delay distance. (After MacCready, 1965 with permission from American Meteorological Society)*

wind direction). Wang (1975), Fritschen and Gay (1979), Brock (1984) and Appendix C provide additional details.

## 5.3 *Deployment of Ground-based Anemometers*

It is necessary to say a few words on the exposure of anemometers and wind vanes since the windfield near the earth's surface varies rapidly with height and is significantly affected by the presence of irregularities in the ground or by nearby obstacles (i.e., trees, buildings). Consequently, the height and conditions for which the measurements are made need to be noted for synoptic reports of surface winds and for general climatological records. The anemometer should be deployed or exposed at an altitude of ≈10 m (33 ft) above the ground and within open surroundings. Open surroundings may be defined as an area where the distance between the anemometer and any obstruction is at least 10 times the height (above the surface) of any obstruction(s). Keep in mind that this is an ideal exposure, and is not likely to be obtainable in practice. However, one should try to choose the site that best approximates this ideal exposure when installing the anemometer and vane. For example, a site on top of a large building or near trees should be avoided if possible. The correct height above the building or other structure required for a representative reading most often exceeds 10 m, but depends on the extent, height, and distance of the obstruction(s) in the vicinity. A rule of thumb for the latter suggests that these measurements should be made at a height above the ground that is at least 2 times the height of the building. A site on a steep hill or the edge of a cliff will be unrepresentative of the general windflow and should not be used, unless of course this is required for the investigation. Anemometers deployed near obstructions yield incorrect mean windspeeds and usually exposes the anemometer to extreme gustiness.

Since it is not always possible to mount the anemometer at the proper height, one may assign it an "effective height". The "effective height" is the estimated anemometer height over flat open land in the vicinity of the anemometer that would

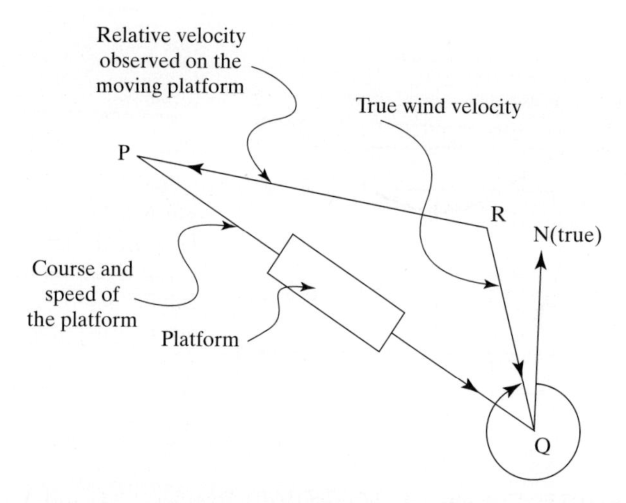


**Fig. 5.10** *The measurement of wind velocity from a moving platform.*

yield the same mean windspeed as an anemometer placed 10 m above the ground given an ideal location. The windspeed at the standard 10 m height based on that measured at some height Z in an obstructed area may be determined by the following expression (BMOAM, 1961)

$$v_{10} = v_Z\{0.233 + [0.656\,\mathrm{LOG}_{10}(Z + 4.75)]\}^{-1}, \tag{5.22}$$

where Z is the height of the anemometer in meters (m), $v_Z$ is windspeed at height Z, and $v_{10}$ is the windspeed at 10 m. The reader is cautioned that this is only an estimate.

There may be one or two exceptions; namely, spaceborne and/or shipborne wind-measuring devices. Spaceborne devices may be free of literally all obstructions, but the instruments used to validate them must represent the true windfield of the region measured from space. All installations aboard a ship, or any other moving platform, indicate the relative measurement of wind velocity and not the true wind velocity. This is illustrated in Fig. 5.10. Here, the segment PQ represents the course and speed of the ship, RQ represents the true wind velocity, and the relative velocity that is observed is represented by RP. Then the direction of the true wind is given by the angle NQR.

We now have a reasonable understanding of how windfield measurements are made by the cup anemometer and standard wind vane, as well as by remote sensing devices. The cup anemometer and wind vane are still the most common means by which to quantify the windfield. However, depending on the particular operational requirements, the recent emphasis on remote sensing techniques, and the reduction in costs for sonic and laser anemometers, these non-traditional windfield devices may soon replace the cup anemometer and standard wind vane. The microwave anemometer may be an exception to the latter, despite the reports of the promising results.

## *Exercises*

1. Modern day cup anemometers are designed with a linear response to windspeed. Explain how the nonlinearity due to overspeeding can cause errors.
2. Show that the time difference equation for the sonic anemometer (Equation 5.13) may be obtained from the acoustic path times of $t_1$ and $t_2$, which are given as

$t_1 = l[(c\cos\Theta) + (u\cos\Theta)]^{-1}$, and $t_2 = l[(c\cos\Theta) - (u\cos\Theta)]^{-1}$ with $\Theta = 0°$, and $u \ll c$.

3. What is the distance constant of the anemometer that produced the record shown in Fig. 5.11 during 4 m s$^{-1}$ winds?

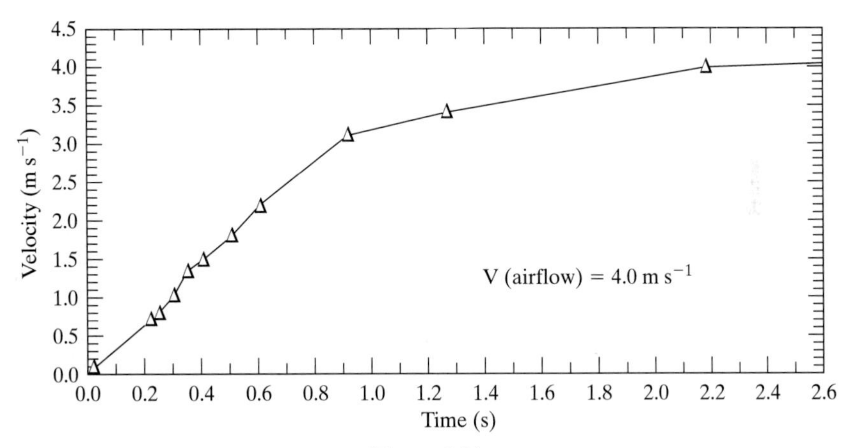


**Figure 5.11**

4. Assuming that the speed of sound, c, may be given by the following equation

   $c = 20.06(T + 273)^{0.5}$,

   where T is the air temperature in °C, Then
   a. Calculate c at $T = -50$ °C and at $T = 50$ °C
   b. Compare this deviation in c with the typical range of windspeed measured by a sonic anemometer.
5. Derive the equation of an underdamped wind vane. State all assumptions and show all steps. Hint: see Appendix C.

CHAPTER SIX

# Water Vapor Measurement

This chapter provides an understanding of how water vapor is measured in the atmosphere via stationary and mobile platforms using present day measuring devices. There is also a description of their time constants and the errors associated with their measurements. Among the present day water vapor measuring devices are remote sensing devices that yield vertical atmospheric profiles of the water vapor content, as well as the total water vapor.

The atmosphere is primarily a mixture of dry air and water vapor. The water vapor in our atmosphere exerts a pressure that is only $\approx$0.01 times that of average sea level air pressure (i.e., 1013.25 mb = 1013.25 hectoPascals, hPa). The water vapor content of the atmosphere is one of the most difficult meteorological parameters to measure. The atmospheric water vapor content varies from virtually none (i.e., water vapor mixing ratio of <0.001 g water vapor per kg of air, or g $kg^{-1}$, at typical atmospheric pressure and temperature conditions) to "completely full" (i.e., air saturated with water vapor, and a saturation mixing ratio as high as $\approx$100 g $kg^{-1}$) over the globe at a given instant. This variability can occur over time periods that are generally shorter than the response times of most humidity measuring devices, even though the modern-day versions are more capable of handling the temporal and spatial variability in water vapor than their predecessors.

The instruments, methods, and procedures for measuring water vapor seem to have, at least in part, emulated from the various ways, both direct and indirect, in which the amount of water vapor in the atmosphere, may be specified. The most common ways to specify the atmospheric moisture content are: vapor pressure (e), absolute humidity ($\rho_v$), specific humidity (q), mixing ratio (w), relative humidity (RH), virtual temperature ($T_v$), dewpoint temperature ($T_d$), and wet-bulb temperature ($T_w$). The hygrometer and psychrometer are two of the most fundamental instruments for measuring the water vapor content in our atmosphere.

A *hygrometer* is the generic name for the device used to measure the amount of water vapor in the atmosphere. This instrument typically converts a variation in relative humidity to a change (relative to an equilibrium condition) in a chemical or physical characteristic of a substance. Examples of physical characteristics of a sub-



106

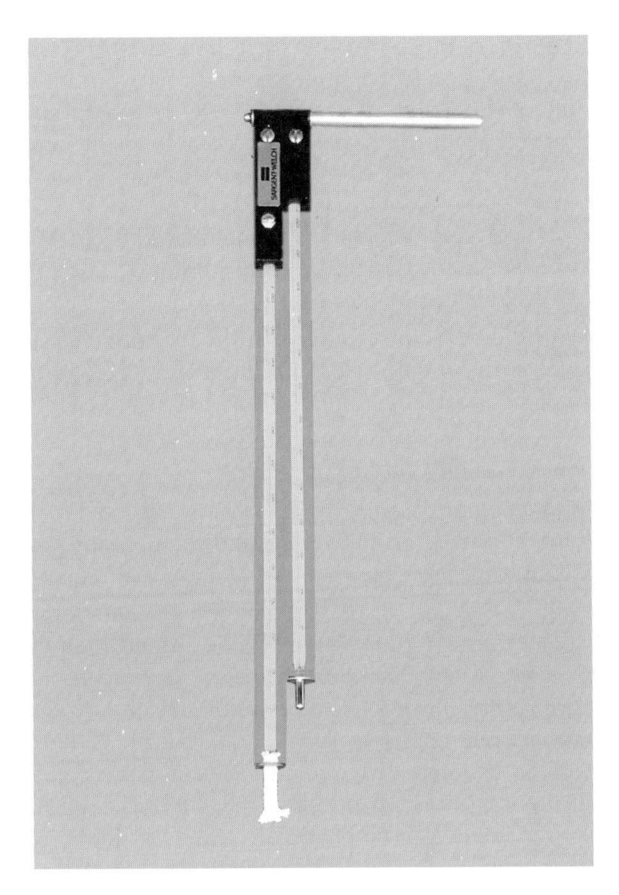

**Fig. 6.1** *A mercury-in-glass psychrometer. (Photograph by J. M. Moran)*

stance include: volume, weight, electrical conductivity, radiation, temperature, water vapor, number of hairs used, the expansion coefficient of the individual hair, and hair strength. (The change in length of hair has been used to measure the amount of water vapor. The response of such an instrument depends on the number of hairs used, the expansion coefficient of those hairs, and the strength of those hairs.)

A *psychrometer* indirectly measures atmospheric humidity via a pair of temperature sensors, typically liquid-in-glass thermometers (Fig. 6.1). The thermometers are identical, except one bulb, or bottom tip of the thermometer, is covered by a muslin sock that is saturated with distilled water prior to a measurement and is referred to as the "wet-bulb." The thermometers are ventilated by swinging them in a circular motion (although fans rotating at calibrated rates are also used). The dry bulb indicates the air temperature, $T_a$, and the wet-bulb indicates the wet-bulb temperature, $T_w$. The resulting difference between these two temperatures is known as the wet-bulb depression which is related to atmospheric humidity. The wet-bulb depression is proportional to the difference between the atmospheric vapor pressure, e, and the vapor pressure, $e_w$, at temperature $T_w$. The proportionality factor, known as the *psychrometric constant,* A, is a function of air ventilation, temperature and density. The derivation of A is detailed by others (e.g., Fritschen and Gay, 1979), and is outlined in Appendix D. The psychrometer yields a relative humidity, RH, within 2–3% of the true value. Note that an error of 0.1 °C results in an error of ≈1% RH. Some practical problems of the psychrometric method for determining the RH include:

1. *Ventilation.* Liquid-in-glass thermometers have a long time constant. Ventilation speeds ideally should be between 4–10 m $s^{-1}$ and constant. Traditional "arm power" does not always provide this.
2. *Heat Conduction.* This would be a concern down at the thermometer stem, and is usually not significant with liquid-in-glass thermometers. Heat conduction errors are more significant if electronic temperature sensors are used.
3. *Lack of Standards.* There are no primary standards (see Glossary) for humidity measurement at this time.
4. *Determination of the actual vapor pressure, e.* If any salts dissolved in the water that the wet-bulb sits in, then these will lower the indicated vapor pressure. A good clean wick and clean water are essential.
5. *Wetness of Wick.* If readings are "unrealistic," namely, the wet-bulb temperature, $T_w$, is equal to the air temperature, $T_a$, and you are not in a cloud, then first check to make sure the wick is wet! Then check for one of the aforementioned problems. One might, as a last resort, check the condensation nuclei (CN) concentrations to make sure there are enough for a visible cloud to form. If there are not enough CN to act as cloud droplet seeds (i.e. insufficient cloud condensation nuclei, CCN), then, in theory, the relative humidity could be ≥100% without seeing a cloud, and the reading might be physically real. The latter would then imply that the wet wick is acting as the site for condensation, thereby heating the wet bulb. Please note that the "last resort" scenario is very unlikely to occur in our atmosphere, especially close to the surface over nearly all of the globe.
6. *Time Constant.* If you are taking measurements while the RH is changing more rapidly then the time constant (i.e., the response of the instrument is slower than the change in humidity), then erroneous measurements will result, as would be true for all instruments.
7. *Slight overestimation of RH* because the $T_w$ is never truly reached. This is often neglected since the magnitude of errors derived from some of the other problems, such as poor ventilation for example, is significantly greater.

The Assman Psychrometer (Fig. 6.2) is an automated version of the psychrometric process. It features a constant windflow (driven by a governor-spring or electric fan) past two thermometers. The thermometers lay inside separate coaxial polished metal tubes. The tube strips the outside layer of air off the bulbs in order to ensure that the instrument is not cooling itself. The bulbs are usually placed inside the tubes at a distance of 1.5 times the diameter of the tube to reduce direct long- and shortwave radiation.

There are other devices for measuring water vapor contents besides the aforementioned ones and they all primarily fall into one of the following four categories:

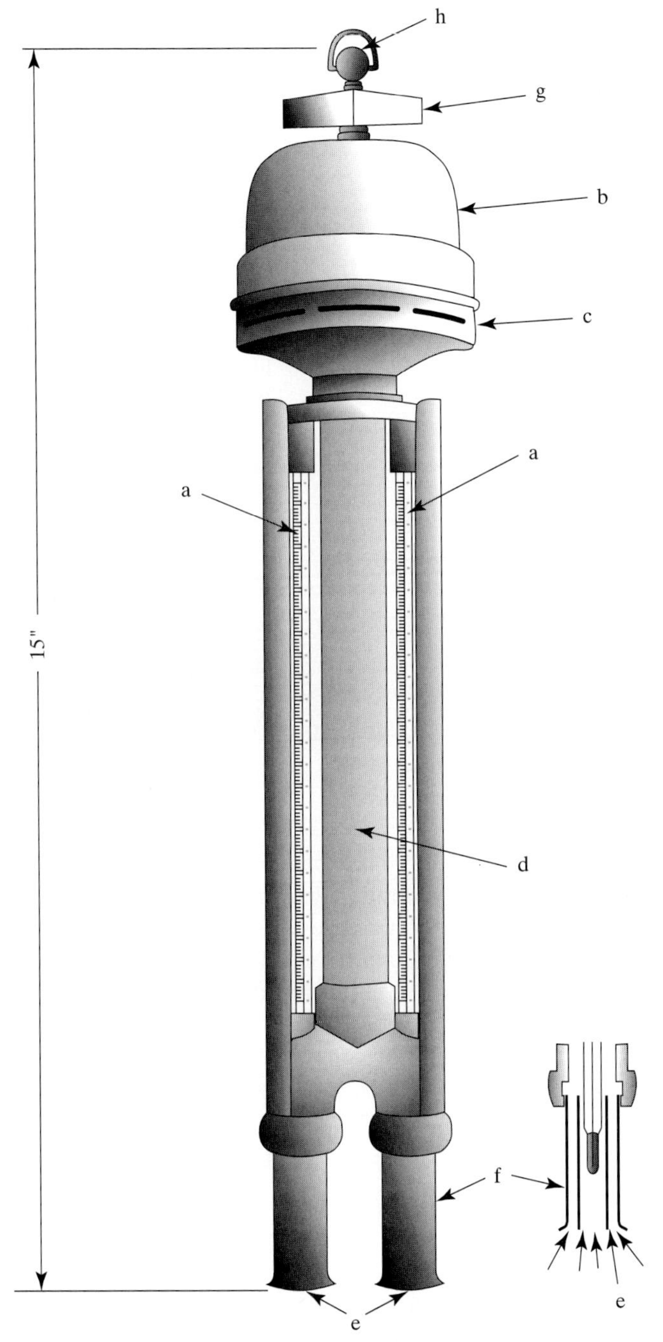


**Fig. 6.2** *An Assmann psychrometer. a: thermometers; b: dome containing the clock mechanism; c: fan and air outlets; d: main air duct; e: air inlets; f: polished tubes protecting thermometers; g: key for winding the clock mechanism (if applicable); h: point of support.(After BMOAM, 1961.)*

**Table 6.1** *Some water vapor measuring devices*

| *Category* | *Method* | *Instrument* |
|---|---|---|
| 1. Addition of water vapor | Non-Electronic | Mercury-in-glass psychrometer |
| | Electronic | Resistance, Thermocouple psychrometers |
| 2. Equilibrium sorption of water vapor by a sensor | Electronic | Electric hygrometers, dew cell |
| 3. Procurement of a vapor–liquid or vapor–solid equilibrium | Dewpoint and frost-point (Electronic) | Dewpoint and frost-point hygrometer, saturation hygrometer |
| 4. Remote sensing devices | | |
| ground-based (possibly airborne) | a. Absorptive | Spectroscopic hygrometer (e.g., ultraviolet, infrared, microwave) |
| | b. Refractive | Interferometer |
| | c. Thermal conductive | Thermal conductivity bridge, absolute humidity recorder, di-electric hygrometer |
| (also spaceborne) | d. Passive, indirect | Radiometer |
| | Passive, direct | Raman lidar |

1. Addition of water vapor to air,
2. Equilibrium sorption of water vapor by a sensor,
3. Procurement of a vapor–liquid or vapor–solid equilibrium, and
4. Remote sensing devices.

Some examples appear in Table 6.1.

Categories 1 through 3, and the ground-based portion of Category 4 in Table 6.1 encompass ≈98% of the instrumentation for measuring humidity. The British Meteorological Office (BMOAM, 1961), Paine and Farrah (1965), Wexler (1970), Wang (1975), Fritschen and Gay (1979), and Brock (1984) provide exhaustive details and plenty of additional references on the multitudes of instruments designed to measure humidity. As a result only a brief description of the most common electronic sensors and the passive and satellite sensors is provided here.

## *6.1. Electronic Devices*

The electronic humidity measuring devices (i.e., hygrometers and electronic psychrometers) consist of thermocouples, resistance thermometers, dew cells, or thin film capacitors to sense a change in the resistance and capacitance of a circuit. This change is then related to the relative humidity. The electronic configuration of the temperature-sensing elements determines whether the device acts as a (1) psychrometer (i.e., adds water vapor to air), (2) a hygrometer whose sensing element absorbs water vapor under equilibrium conditions, or (3) a hygrometer that directly or indirectly establishes the amount of water vapor in the air by changing a physical

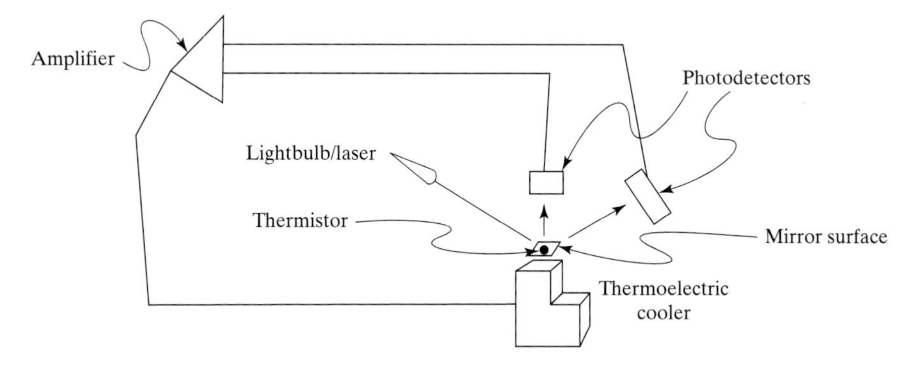


**Fig. 6.3** *A schematic of the generic dewpoint hygrometer.*

quantity (such as temperature of the air, conductivity, capacitance, or resistance of the sensor). The device in (1) contains two electronic temperature measuring devices, one of which is covered by a wet sock (usually muslin) and ventilated. The resulting change in the resistance is compared to a similar sensor that is also ventilated but kept dry, much like the traditional liquid-in-glass psychrometer. Two thermopiles, or similarly configured resistance thermometers, with the sensing junctions of one kept moist and the other dry are used as the temperature measuring devices. A platinum wire ($\approx$0.1–$\approx$0.5 mm diameter) resistance thermometer arranged in a helix around a strand of wet cotton wicking with either platinum or nickel as the dry element yields a response time of $\approx$0.2 to 1.0 s in the wet-bulb temperature, $T_w$, range of 0 to 20 °C (e.g., McIllroy, 1955, 1961).

The dewpoint hygrometer (Fig. 6.3) is used to measure $T_d$, and is very capable of measuring low or very high water vapor concentrations without measuring water vapor itself. The theory behind this technique is briefly stated as follows. Air with a given temperature, $T_a$, pressure, P, and mixing ratio, w, is cooled until it reaches its saturation point with respect to a free water surface. The temperature of the saturation point is technically termed the *thermodynamic dewpoint temperature,* $T_d$. (If the temperature falls below freezing during its cooling, and the saturation point with respect to ice is reached, then ice is formed, and that temperature is the *thermodynamic frost-point temperature,* $T_f$.) The saturated vapor pressure with respect to water, $e_s$, is a function of T and is mathematically expressed by the empirical formula (Rogers and Yau, 1991),

$$e_s(T) = e_s(T_0)\exp[(17.67T)(243.5 + T)^{-1}] \tag{6.1}$$

where T is in degrees C, $T_0$ is a reference temperature (usually 0 °C), and $e_s(T_0)$ is 6.112 mb, over the temperature range $-30\,°C \leq T \leq 35\,°C$ with a fit that is within 0.1%.

The dewpoint hygrometer consists of three general components: sensor, optical detection, and thermal control modules. The sensor module consists of a thin metallic mirror—about 1.5 mm in diameter, that is thermally regulated by a cooling module—and a temperature measuring device (such as a thermistor, thermocouple, resistance thermometer, or capacitive element) imbedded beneath the mirror. The mirror material (i.e., silver, copper, gold, stainless steel, and their alloys) has a very high thermal diffusivity in order for its surface temperature to respond rapidly to the thermal regulation system. The higher the thermal diffusivity, the smaller the temperature gradient between the mirror's surface and the thermal sensor, and the greater the precision of the measurement. The mirror has to be free of atmospheric contaminants, since they will alter the vapor pressure (water soluble matter tends to lower the vapor pressure), and hence the dewpoint of the condensate. An electro-optical system is employed to accurately determine the dewpoint temperature of moist air by monitoring the thickness of the condensed deposit using an incandescent light and photodetector. The thickness of the liquid water deposit is preserved by controlling the evaporation/condensation rate on the surface (i.e., a servo-controlled cooling/heating module). The temperature of the mirror is adjusted such that the vapor pressure over the liquid deposit just equals the vapor pressure of the air, and a quasi-steady state equilibrium exists. At this equilibrium point, the mirror temperature equals the dewpoint temperature. The optical detection module houses a narrow beam of incandescent light with an incident angle of $\approx 55°$ directed at the mirror. The intensity of the reflected light, or reflectance, is monitored by a photodetector, that, in turn regulates the cooling/heating module via a servo-control. See Wang (1975) and Fritschen and Gay (1979) for additional details. A major shortcoming of this hygrometer, and a starting point for future research, is the uncertainty in knowing when the dewpoint temperature is "truly" reached. Gerber (1980) describes and has used a variation of the dewpoint hygrometer, termed the saturation hygrometer, with very reasonable success. The saturation hygrometer is capable of precisely measuring the relative humidity between 95 and 105%. Note that a frost-point hygrometer measures the frost point, $T_f$, in a similar fashion.

The present day dew-cell hygrometer (Fig. 6.4) consists of a bifilar heater composed of a pair of gold wire electrodes that are wound around a fiberglass or dacron wick saturated with lithium chloride (LiCl), which encompasses a tube that contains a temperature sensor. Older versions have used a silver, rhodium plated silver, or palladium wire instead of the gold treated with polyvinyl acetate if the LiCl was not available. Lead wires attached to an aluminum rod and a conducting layer have also been used in the past with limited success near saturation.

The dew-cell hygrometer works on the principle of temperature-controlled saturated salt solutions. An alternating current is connected to the bifilar winding through a current-limiting resistor. When the wick is wet, the resistance between the wires is low and current will flow causing the element to heat. This, in turn, causes water to evaporate thereby increasing the resistance and reducing the current. Ultimately an equilibrium temperature is achieved wherein the vapor pressure of the lithium chloride solution is in equilibrium with the vapor pressure of the air. The atmospheric dewpoint temperature is a linear function of the dewpoint temperature of the lithium chloride, or what has been called the bobbin temperature. The RH range of the dew cell is limited from 11 to 100% at room temperature. The sensor must be constantly powered to prevent the LiCl salt from absorbing moisture and dripping off. Note that since LiCl polarizes, AC

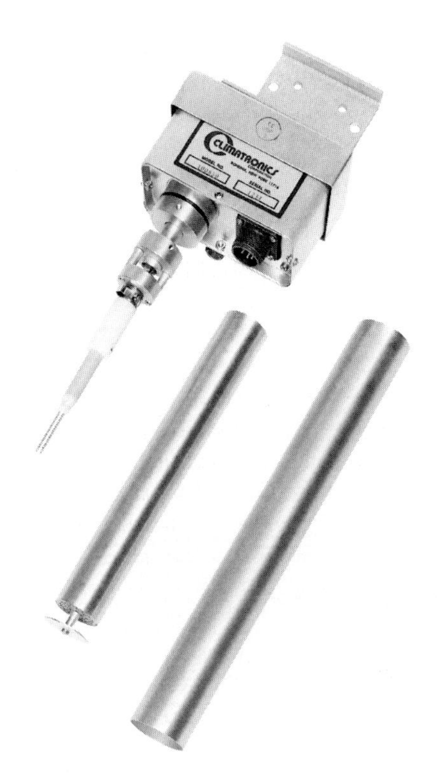

**Fig. 6.4** *A photo of a modern LiCl dew-cell hygrometer, and naturally aspirated shield. (Courtesy of D. Gilmore and D. Katz of Climatronics Corp.)*

power is needed. The sensor may be retreated without disturbing the fundamental calibration. The time constant of a dew cell is about 5 min., with an accuracy in dewpoint measurement of $\approx\pm1.5\,°C$. Since the operating temperature of the dew cell is well above the ambient temperature, care must be taken to minimize heat loss by convection, radiation, and conduction.

There is also a humidity sensor that operates on a capacitance change of a polymer thin film capacitor developed in Finland by Vaisala; additional details appear in Chapter 8. Simply, a thin gold resistance grid is sandwiched between two polymer layers $\approx 1\ \mu m$ thick. The absorption of water vapor by the polymer alters its capacitance. The material of the polymer is resistant to most chemicals and the calibration is not affected by liquid water. The humidity range is 0 to 100%, and it responds to a 90% humidity change in less than 1 second. Its accuracy is better than $\pm2\%$.

A wet-bulb psychrometer is usually used as a calibration standard for humidity sensors, primarily because it is based on the hygrometric equations, which are based on the comparison of many psychrometric readings with either absorption or dewpoint hygrometers. Otherwise there is no primary standard. A dewpoint hygrometer does not require a calibration in the sense of comparison with standards, but they should be compared against another dewpoint hygrometer. A very simple way of calibrating the water vapor measuring devices is to place and operate them in a humidity controlled mixing chamber wherein the air can obtain a homogenous humidity with trays of water and salt solutions (e.g., zinc chloride, lithium chloride, calcium chloride, sodium chloride, potassium chloride, sodium carbonate ... ) at its bottom. Perhaps radiometers will become the primary standards.

## 6.2. *Remote Sensing Devices*

The remote sensing devices usually report their water vapor information in terms of a mixing ratio, precipitable water or an equivalent term, but not in terms of percent relative humidity. Some of them, like the electronic moisture measuring devices, will report their water vapor information in terms of a temperature leaving the individual to convert the temperature into relative humidity. The latter conversion can cause additional errors and hence a loss of accuracy; let alone resolution in the water vapor information compared to more direct methods. Nonetheless, the remote sensing devices highlighted below are arbitrarily separated into ground-based and spaceborne categories, even though many, if not all, of the devices mentioned could probably be adapted to airborne platforms.

***6.2.1. Primarily Ground-based.*** Infrared and microwave hygrometers are based on the principle that water absorbs energy at certain wavelengths and not others (e.g. Wood, 1963). Sensors that absorb radiant energy are essentially radiometers. Guirard et al. (1979) used a ground-based dual channel microwave radiometer system to determine the precipitable water that had one radiometer operating at a wavelength of 1.45 cm (i.e. frequency of 20.6 GHz) and the other at 0.95 cm (31.6 GHz). They chose the 1.45 cm absorption band because of the absence of a pressure broadening, or, in other words, an absorption band whose signal width will not be dependent on the slight change in the energy levels of electrons and in the energy of an absorbed photon. The 0.95 cm wavelength was chosen because it is in the window between the water vapor and oxygen absorption lines. Comparing their measurements to precipitable water estimates from radiosondes they are able to retrieve a precipitable water that is within 0.01 cm. Radiometrics Corp. (1995) makes a dual wavelength microwave water vapor radiometer at frequencies of 23.8 GHz and 31.4 GHz that weighs 15 kg, automatically checks its calibration every 10 s, consumes a relatively low to moderate amount of power (17 W to 30 W during elevation scans), has a 1 s time constant, and has a $\pm 0.5\,^{\circ}K$ sky brightness temperature accuracy.

A Lyman-alpha hygrometer output is high-pass filtered at a frequency, $f_l$, and low-pass filtered at a frequency, $f_u$, and the difference between the $f_l$ and $f_u$ outputs is proportional to the measured water vapor variance in the bandpass region. The resulting wavelength is at 0.12156 $\mu$m and is termed the Lyman-alpha line of the hydrogen spectrum. Near this wavelength oxygen absorption is minimized (absorption coefficient $\approx$0.3), and the absorption coefficient for water vapor is 0.0387 $\mu m^{-1}$.

Consequently, the Lyman-alpha hygrometer operates by passing Lyman-alpha radiation through a sample of air and measures the incident and transmitted radiances from which the relative humidity of the air may be calculated. The path lengths may be a few cm or less and the time response is nearly instantaneous. See Fleagle and Businger (1980) and Brock (1984) for details.

A Raman lidar has been used to obtain a profile of atmospheric water vapor (Goldsmith et al., 1994). The Raman lidar sends out a laser pulse and records the atmospheric backscatterred return signal as a function of time as in most lidar or radar systems (Chapter 2). The return signal consists of elastic and inelastic scattered components, the latter of which is used to provide profiles of water vapor. The inelastic scattering is the result of the vibrational Raman effect that shifts the incident wavelength by a frequency characteristic of the molecule (i.e., 0.3652 $\mu m^{-1}$ for water vapor). The return signal also provides information about nitrogen which has a frequency characteristic of 0.2331 $\mu m^{-1}$. The water vapor mixing ratio as a function of altitude may be obtained from the ratio of the Raman lidar water vapor signal to the nitrogen signal, and only requires a small correction for differential attenuation at the two return wavelengths (Whiteman et al., 1992). However, the relative humidity also requires that a co-located temperature profile measurement is made. Given the latter, then the root mean square error in the mixing ratio profiles derived from a Raman lidar are less than 0.15 g $kg^{-1}$. However, at present the Raman lidar is essentially only operable during nighttime hours. Goldsmith et al. (1994) has additional details.

The spectral radiance experiment, SPECTRE (Ellingson and Wiscombe, 1996), is a surface-based system to remotely profile temperature, humidity, aerosol, and cloud. The profiles are blind to variations below 100–300 m, and its instrumentation are presented in Table 6.2. The results are promising, but the system they discuss is still experimental.

***6.2.2. Primarily Spaceborne.*** Satellite water vapor retrieval is routinely done by the National Weather Service for operational purposes. The development of

**Table 6.2** *Remote profiling instrumentation used in SPECTRE (after Ellingson and Wiscombe, 1996)*

| *Instrument* | *Measured quantity* | *Vertical range (km)* | *Vertical resolution (m)* | *Averaging time (min.)* | *Error* |
|---|---|---|---|---|---|
| Raman lidar | water vapor mixing ratio | 0.3 to 7 | 100–150 | 2 | 0.1–0.2 g $kg^{-1}$ or 5% |
| Radio Acoustic Sounding System (RASS)[a] [400-MHz] | virtual temperature | 0.3 to 3 | 150 | 10 | 0.5–1.0 °C |
| RASS [50-MHz] | virtual temperature | 2 to 7 | 150 | 10 | 0.5–1.0 °C |

[a]See May et al. (1988) for details. The RASS uses radar signals backscattered and Doppler shifted by acoustic waves to determine the sound speed and subsequently the virtual temperature.

improved water vapor profile retrieval algorithms from satellites also continues. Satellite-derived water vapor values are retrieved on the principle that water vapor emits radiation at certain wavelengths. Radiation received by a satellite sensor (simply a radiometer) is electronically filtered so that only radiation from the specified wavelengths is stored. This "stored" signal is essentially an electrical resistance that is converted into a temperature, termed the *antenna temperature.* The antenna temperature may then be converted to a *brightness temperature,* $T_b$ since

$$T_b \alpha \, L T_a \tag{6.2}$$

where $T_a$ is the antenna temperature, and L is the radiance of the water vapor ($H_2O_v$) molecules in the path from the satellite to a surface (typically earth), or infinity as discussed in Chapter 2. The radiance received depends on the number density of the $H_2O_v$ molecules. The satellite sensors for water vapor consist of the microwave limb sounder (MLS), microwave sounding unit (MSU), second generation high resolution infrared radiation sensors (HIRS/2), special sensor microwave imager (SSM/I) and in the near future a Raman lidar (see p. 115).

The SSM/I, MSU, MLS and HIRS/2 are all very thoroughly described by Kidder and Vonder Haar (1995). The radiance measurements from the HIRS/2 and MSU are primarily centered on the carbon dioxide, $CO_2$, and ozone, $O_3$, absorption bands which are used to simultaneously generate temperature and moisture soundings. The retrieval of moisture information, especially, using infrared, IR, wavelengths is not very accurate, since the retrieval algorithms yield little or no information on atmospheric water vapor and can cause a "dry bias" (Wu et al., 1993). In contrast, Stephens et al. (1994) show that limitations in IR radiative transfer theory can cause a significant overestimation of water vapor in regions of large scale subsidence. As a result, SSM/I data are given a higher total column water vapor confidence level then the combined HIRS/2 and MSU data (Randel et al., 1996). Nonetheless, keep in mind that SSM/I data yield large overestimates of the water vapor content when retrieved over land and sea ice surfaces. The combined data from SSM/I, HIRS/2 and MSU data yield average total column water vapor for the northern and southern hemispheres of $\approx$26 ($\pm$3) mm and $\approx$23 ($\pm$2) mm during the period between 1988–1992. The latter combination of data used to retrieve the total column water vapor, in this case, is known as a multispectral retrieval technique.

The MLS has been in operation since 1991 and measures radiation near a wavelength of 1.5 mm (frequency $\approx$205 GHz), which primarily picks up the emission from water vapor in the upper troposphere. The MLS measurements are strong functions of temperature. The instrumental noise on each 2-s integration for the 1.5 mm channel is $<$0.1 °K equivalent brightness temperature. Read et al. (1995) provide additional details of the MLS and show that it can provide information on global upper tropospheric water vapor. The algorithm for retrieving the water vapor continues to be developed and a next generation MLS is being developed for NASA's Earth Observing Systems.

## 6.3. *Response Time and Spatial Resolution*

The time it takes for a hygrometer to respond to 63.2% of a step change in the water vapor content (i.e., its time constant) is mathematically described as follows. If $\lambda$ is the time required to bring the initial indication of the ambient air humidity, $RH_o$ (at $t = 0$) to 63.2% of the final indication of the ambient air humidity, $RH_f$, then the ambient air humidity reading $RH_t$ (after t seconds) may be written

$$RH_f - RH_t = \lambda d(RH_t) dt^{-1} \quad \text{and} \quad \Delta RH = \Delta RH_0 e^{(-t\lambda^{-1})}, \tag{6.3}$$

where $\Delta RH_o$ is the increment of humidity of the immediately preceding step function, and the rest follows similarly to temperature. Time responses for various types of hygrometers are shown in Table 6.3.

We now have a better understanding of the operation, response, and the accuracies of the water vapor measuring devices in our atmosphere. The remote sensing devices, especially the microwave radiometer, generally appear to be quicker and at least as accurate as the other water vapor measuring devices that are presently available. The data from the radiometers may also be used to retrieve the vertical profile of water vapor instead of simply an in-situ value. However, please recall our discussion at the end of Chapter 3, and keep in mind that the retrieval of water vapor information from a radiometer, whether ground-based or especially satellite-based, is highly dependent on modeling efforts. The modeling efforts are in turn strongly dependent on the scheme used to validate the model.

**Table 6.3** *The time constant of some hygrometers*

| *Type* | *Time Constant (s)* | *Accuracy* |
|---|---|---|
| Psychrometer (Liquid-in-Glass) | 300 | $\pm 4\%$ to $\pm 15\%$[a] for $10\% \leq RH \leq 100\%$ |
| Dew cell | 10–500 | $\pm 2\%$ for $RH \leq 80\%$, $\pm 3\%$ $RH > 80\%$ |
| Capacitive | 10–30 | $\pm 2\%$ for $RH \leq 80\%$, $\pm 3$–$5\%$ $RH > 80\%$[b] |
| Dewpoint | <2–10 | $\pm 0.5\,°C$ |
| Microwave radiometer | 1 | $\pm 0.5\,°K$ brightness temperature |

[a]Higher end expected in rainy or freezing weather. Accuracy is also reduced for $RH < 10\%$

[b]Handar carries a sensor (435C) that has $\pm 2\%$ for $RH \leq 100\%$.

## *Exercises*

1. Is moist air less dense than dry air? Why?
2. A psychrometer uses the latent heat of vaporization to cool the wet-bulb thermometer. That is, water vapor molecules go from the wet-bulb to the air. Consider the effect of wetting the wick on the "wet-bulb" thermometer with a solution that absorbs water vapor. Will the wet-bulb be cool or warm? What are the advantages or disadvantages of such a system?
3. Most manufacturers of electronic water vapor measuring devices report the accuracy of the dewpoint or "wet-bulb" temperature. Calculate the equivalent accuracy in relative

humidity as a function of temperature for a dewpoint error of 0.1 °C and 1 °C. Use a temperature range of −5 °C to 25 °C. Be careful below freezing.

4. Make humidity measurements at various times (i.e., 0, 0.5, 1, 2, 5, 10 and 15 minutes) with a liquid-in-glass psychrometer and an electronic psychrometer or another ground-based moisture measuring device. Repeat at a second location. Keep the psychrometers covered until measurements begin; write down the initial time and readings as soon as you take the instrument out of its cover. Then note subsequent readings and the general weather-related observations; especially if there are any changes that might affect your readings (i.e., the sky was overcast and then became clear during your measurements). Be careful not to introduce any systematic or random errors. Locations might include one or more of the following;
   a. Grass–Shade
   b. Grass–Sun
   c. Concrete–Shade
   d. Concrete–Sun
      (1) Obtain seven or eight measurements of T and $T_w$ at each location.
      (2) Calculate the relative humidity.
      (3) Estimate the time constant using (a) the graphical technique and (b) the relevant theory. Discuss your findings, including all possible errors.

5. Using your data in Exercise 4 as the column averaged values, estimate the water vapor mixing ratio in that column using one of the satellite sensors. Discuss your choice of satellite sensor. Justify any assumptions you need to make.

6. A psychrometer is ventilated at 5 m s$^{-1}$. The albedo of the dry bulb is 0.9, and that of the wet bulb is 0.4. Calculate the error in wet-bulb depression and relative humidity if the sensors are unsheltered and exposed to an insolation of 1,000 W m$^{-2}$. Assume the sensors are cylinders of 0.5 cm diameter and 3 cm in length.

CHAPTER 7

# Cloud And Precipitation Measurement

This chapter is primarily concerned with cloud and precipitation measurements, the principles behind such measurements, and how well such measurements are made.

## *7.1. Clouds*

***7.1.1. Background.*** Aerosol particles with diameters between $\approx 0.05$ and 5.0 $\mu$m make up a significant portion of the atmospheric particles within our atmosphere, and they are either very poor at absorbing water vapor (hydrophobic), "nonchalant" about absorbing water vapor (hydrophilic), or very good at absorbing water vapor (water soluble or hygroscopic). Actually, the hydrophobic particles generally make the best "seeds" upon which ice crystals grow, and the hygroscopic particles make the best "seeds" upon which water droplets grow. The backscattering and total-scattering properties of aerosols with diameters between 0.05 and 5 $\mu$m are measured by a Nepholometer (e.g., a TSI nephelometer-model 3563; see Fig. 7.1), and are used to infer information concerning (i) physical and chemical characteristics, (ii) the radiative effect, and (iii) the validation of remote sensing retrievals of physical and chemical characteristics (i.e., concentration, size spectra, bulk chemical composition, etc.) of aerosols in this size range. For example, Hindman et al. (1985) have suggested a dual wavelength (i.e., near infrared and red-visible) remote sensing technique for retrieving simple characteristics of haze layer aerosols with 1 to 6 $\mu$m diameters. Research relevant to the remote sensing of aerosols continues, and Lodge (1994) has put together an excellent major reference for aerosol-measuring techniques.

Clouds form on a subset of hygroscopic aerosol particles, known as cloud condensation nuclei or CCN, as the air in which they reside is brought to its saturation point. Air is brought to saturation by one or more of the following processes: radiative cooling, evaporation-mixing, moist air advection into cooler air, and lifting (such as frontal boundary, surface convergence—not necessarily confined to the surface of the earth—and orographic lift). The processes involved

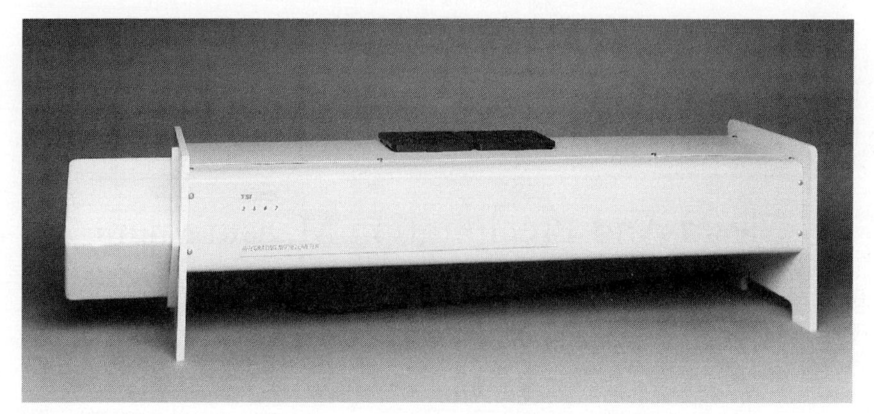

**Fig. 7.1** *A photo of a TSI-3563 nephelometer. (Courtesy of TSI Incorporated.)*

in cloud formation, development, and decay are complicated (e.g., Cheng, 1970; Schaefer and Cheng, 1971; Wallace and Hobbs, 1977; Rogers and Yau, 1991; Pruppacher and Klett, 1980; Houze, 1993; Young, 1993); modelled (e.g., Cotton and Anthes, 1989; Orville, 1996); simulated (e.g., Grantham, 1995); and studied via remote sensing techniques (e.g., Battan, 1973; Kidder and Vonder Haar, 1995; others). DeFelice and Cheng (1998) discuss the aerosol-cloud interactions at the end of a cloud cycle, particularly during non-classical occurrences (Fig. 7.2). The underlying purpose of their discussion was to provide a starting point for future research which is visualized in Fig. 7.3. Normally, or "classically," the number of particles observed after evaporation is similar to that at the onset of the evaporation, unless the evaporation marks the boundary between two air masses (i.e. front or upper-level disturbance). Air originating over a continent typically contains more particles than air originating over the ocean. Nonetheless, there are occasions when the environmental relative humidity is below the point required to crystallize the most soluble salt component of the evaporating drop, and the environment into which the cloud droplets evaporate is conducive to new particle production, or non-classical nuclei generation.

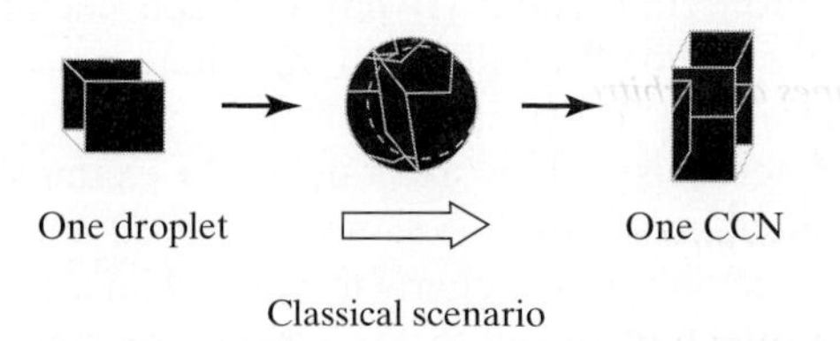


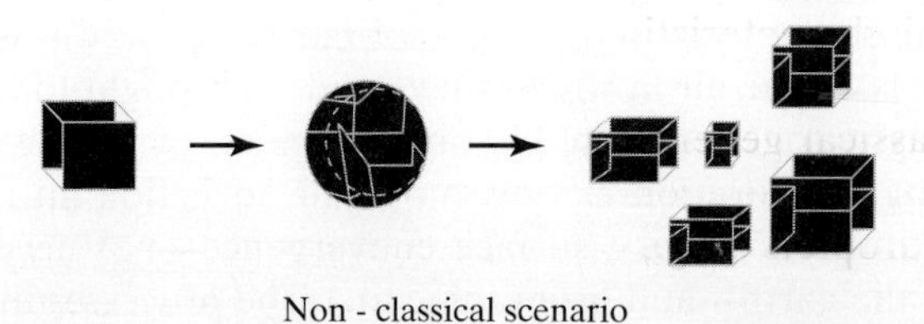


**Fig. 7.2** *A conceptualization of the "classical" versus "non-classical" yield of CCN upon the evaporation of a cloud droplet. Note that the shapes are arbitrary.*

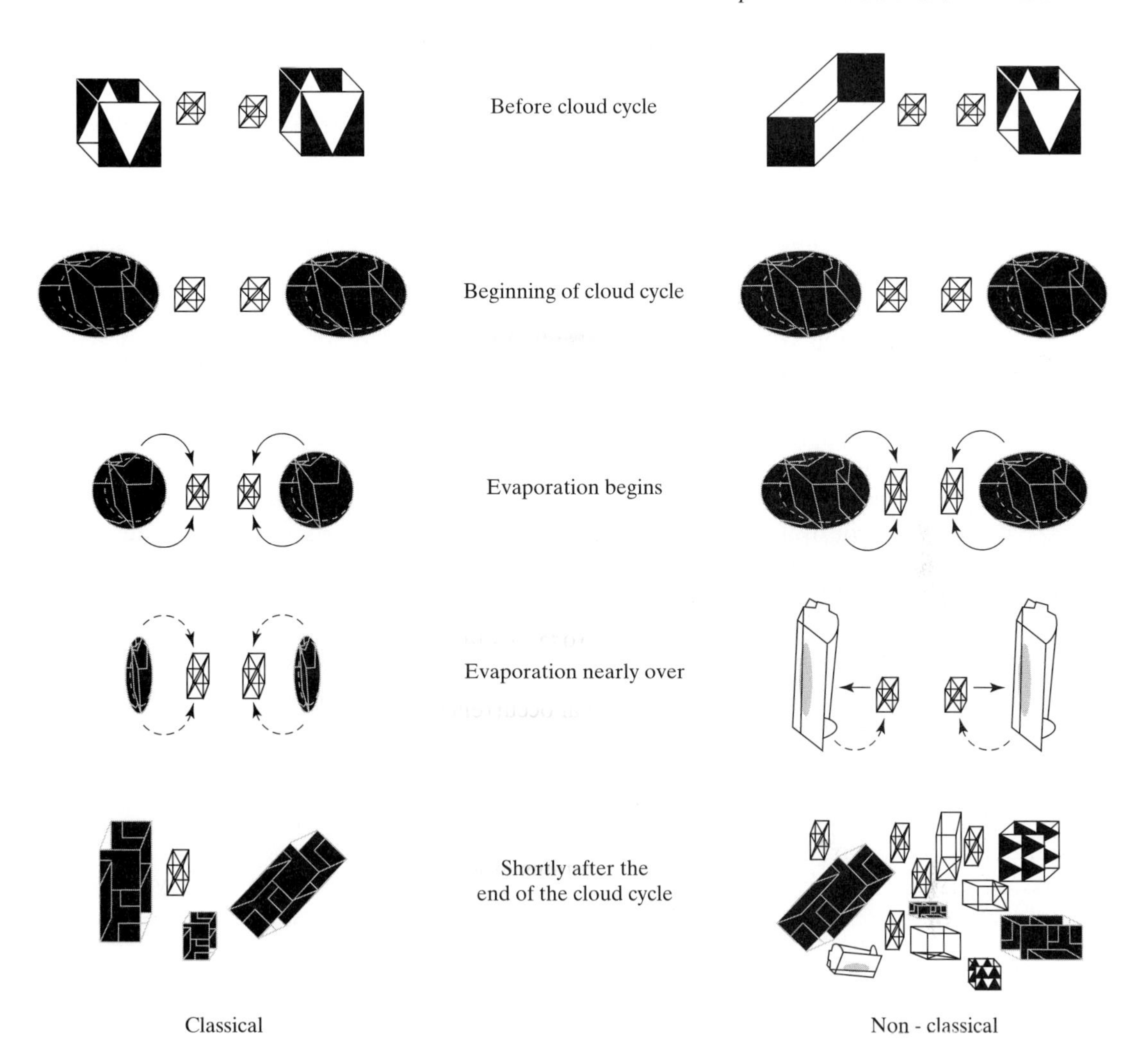


**Fig. 7.3** *A preliminary mechanism to explain the intermittent occurrence of the non-classical generation of nuclei at the end of a cloud cycle. Nuclei shapes are arbitrarily drawn and are dependent on the chemical composition. The "clear" unfilled nuclei are hollow. The arrows indicate vapor flux, with dashed arrows suggestive of a smaller flux. The gray shading denotes the presence of a liquid phase (not necessarily water) within the central portion of the emerging hollow nuclei.*

At this point the exact mechanism has not been clearly established, although observations suggest the generation has a fundamental dependence on temporal and spatial timing, sampling frequency, as well as the meteorological, physical, and chemical characteristics of the cloud droplets and their surroundings. A proposed preliminary mechanism to explain the intermittent occurrence of the non-classical generation of nuclei at the end of the cloud cycle is illustrated in Fig. 7.3. Briefly, hollow nuclei emerge from the originally evaporating sulfate-rich droplets (e.g., RH $\approx$ 79%), and fragment (i.e., the emergent particle or nuclei is hollow and breaks apart) as the surrounding aerosols, which have grown since the evaporation began now evaporate some of their vapor, since the air is still apparently subsaturated. The fragments may also grow some as long as an equilibrium state has not been established, thereby leading to a non-classical generation of nuclei. Appendix E provides additional details, and outlines an experimentalist's investigation of this phenomenon.

The aforementioned greater than expected nuclei emerging at the end of a cloud cycle is important for at least two simple reasons: (1) the production of uncommonly high nuclei concentrations is a scientific reality—albeit an apparently infrequent reality—that might be climatically important (e.g., Twomey et al., 1984; Charlson et al., 1987; Ghan et al., 1990), and (2) it reminds us that naturally occurring phenomena do not always occur when and how we want them to, or think they should for that matter. The second reason is one of the primary incentives for applying the principles outlined in Chapter 1 whenever measurements are required. If an experimental program to investigate any natural phenomena (i) is planned for the "smallest common denominator" affecting that phenomena (of course, bearing in mind all constraints of the particular program), (ii) follows a prudent standardized QA plan, and (iii) is allowed to exist for at least 5 years, then any phenomena, or part thereof, will become very well understood. The more we understand a phenomena the better we can predict it, and the better we can predict something, the easier it will be to ensure that its effect on the environment will be beneficial to all.

The growth of cloud droplets from vapor may be expressed by the diffusional growth equation, assuming (a) the drop is a sphere and has a density of pure water (see Saxena and Fisher, 1984 for the validity of the latter assumption), (b) all vapor and thermal gradients near the surface of the drop are uniform, (c) kinetic, ventilation, nonstationary growth, nonsteady updraft and statistical effects are negligible (e.g., Rogers and Yau, 1991), and (d) the temperature and vapor density at the surface are approximately equal to the air temperature and the saturated water vapor density, respectively. Then the expression for growth by condensation in terms of radius and saturation ratio becomes

$$r[dr\,dt^{-1}] = [S - 1 - (ar^{-1}) + (br^{-3})][F_T + F_D]^{-1}, \tag{7.1a}$$

where S is the saturation ratio or actual vapor pressure divided by the saturation vapor pressure at air temperature, T, pressure, P, and is defined as

$$S \equiv 1 + (ar^{-1}) - (br^{-3}) \tag{7.1b}$$

$(ar^{-1})$ represents the curvature effect which expresses the increase in saturation ratio over a droplet compared to a plane water surface, and "a" may be approximately expressed as

$$a = 2\sigma_{la}(R_v \rho_v T)^{-1} \approx 3.3 \times 10^{-5}(\text{cm}\,^{\circ}\text{K})T^{-1}, \tag{7.1c}$$

$\sigma_{la}$ is the surface tension of the droplet at its surface–air interface, $R_v$ is the gas constant for water vapor, $\rho_v$ is the density of water vapor; $(br^{-3})$ represents the solution effect which expresses the lowering of the saturation ratio due to the presence of a dissolved substance of molecular weight, $m_s$, in the amount, m, having i components (e.g., NaCl has i ≈ 2; ammonium sulfate has i ≈ 3), and may be expressed as

$$b \approx 4.3(\text{cm}^3)\,\text{i}\,\text{m}\,\text{m}_s^{-1}; \tag{7.1d}$$

and $F_T$ and $F_D$ are the thermodynamic terms associated with heat conduction and vapor diffusion respectively. Rogers and Yau (1991) provide tables to obtain the values of $F_T$ and $F_D$ given an air temperature and pressure. For example, at 0 °C and 900 mb, $F_T + F_D \approx 1.58489 \times 10^{-6}$ (s cm$^{-2}$) and at −4 °C and 900 mb, $F_T + F_D \approx 1.99526 \times 10^{-6}$ (s cm$^{-2}$).

The measurements of clouds often entail the quantification of their liquid water content (and ice content if appropriate), the frequency distribution of droplet sizes within the clouds, the chemical composition of the cloud droplets (e.g., Hegg and Hobbs, 1979, 1986; Lin and Saxena, 1991), the meteorological characteristics associated with the cloud, and the interstitial components within the cloud, or the components that exist between the cloud droplets (e.g., Hudson, 1984; Ahr, Flossmann and Pruppacher, 1989; DeFelice and Saxena, 1994). Hoppel (1988), Hoppel et al. (1990), and others discuss the interactions between the interstitial nuclei and the droplets.

The liquid water content of a cloud (LWC) has units of g m$^{-3}$ and is defined by

$$\text{LWC} = (4\pi/3)\int_0^\infty \rho N(r)\, r^3\, dr \tag{7.2a}$$

or

$$\text{LWC} \approx 0.5236\int_0^\infty \rho N(D)\, D^3\, dD, \tag{7.2b}$$

where $\rho$ is the density of water, N(r) is the number density of droplets with radius r, and N(D) is the number density of droplets with diameter D. Equation 7.2 shows us that the liquid water content is a product of two variables, and this implies that two clouds with different droplet sizes may yield the same liquid water amount.

The LWC is actually very easily measured and is operationally used for cloud-seeding activities for example. Demoz et al. (1993) used a mobile ground-based spinning reflector radiometer to measure liquid water content and vapor content in Sierra Nevada storms with good results. Other simple ways to measure LWC fall under a grouping known as gravimetric methods. These include, for example, a radiosonde modified to detect supercooled water amount (Hill, 1990); an improved "cloud gun" (Hindman, 1987); or a bulk technique first developed for cloud chemistry studies (by the late Bill Winters in the late 1970s at Atmospheric Science Research Center, ASRC, of the State University of New York at Albany).

The Atmospheric Science Research Center, ASRC, passive string cloudwater collector (Winters et al., 1979) is not the only device used to obtain cloudwater samples. There are a few other kinds of cloudwater collectors, such as, one to measure fogwater (Jacob et al., 1985); the Desert Research Institute (of the University of Nevada, Reno) linear jet impactor; and the DRI (Katz and Miller, 1984) and the California Institute of Technology (Caltech) active strand cloudwater collectors

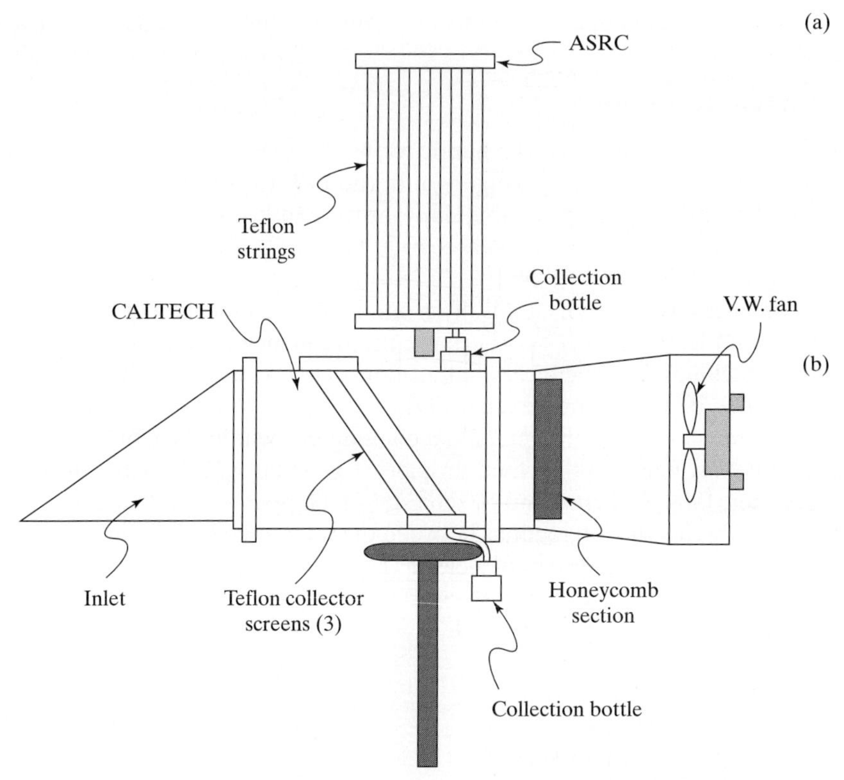


**Fig 7.4** *Examples of some cloudwater collectors. (a) The passive string ASRC cloudwater collector. (b) The Daube Caltech active strand cloudwater collector (CALTECH). (c) The DRI cloudwater collector. (d) The Caltech active strand size fractionating cloudwater collector. (Source provided in text and with permission from Elsevier Publishing Company.)*

(Daube et al., 1987; Demoz et al., 1996). Figure 7.4 shows an example of some of the cloudwater collectors. There have been a number of cloudwater collector intercomparisons (e.g., Gertler, 1983; Hering et al., 1987; DeFelice and Saxena, 1990; Demoz et al., 1996).

The ASRC is useful in most cases, except when the windspeed is less than $\approx 3$ m s$^{-1}$ and greater than $\approx 19$ m s$^{-1}$. The collection efficiency of the ASRC collector is maximum for typical cloud droplet sizes, and is reported as being dependent on windspeed (e.g., Winters et al., 1979; McLaren et al., 1985). Actually, the collection efficiency of the collector increases with time once collection begins, and it becomes independent of windspeed once the collected cloudwater or precipitation water begins falling into the collection bottle (DeFelice and Saxena, 1990). The ASRC collector obtains a bulk water sample consisting of cloud droplets plus small precipitation size hydrometeors if present. If you require size-differentiated cloudwater chemistry then the Caltech active strand size-fractionating cloudwater collector could be used. It

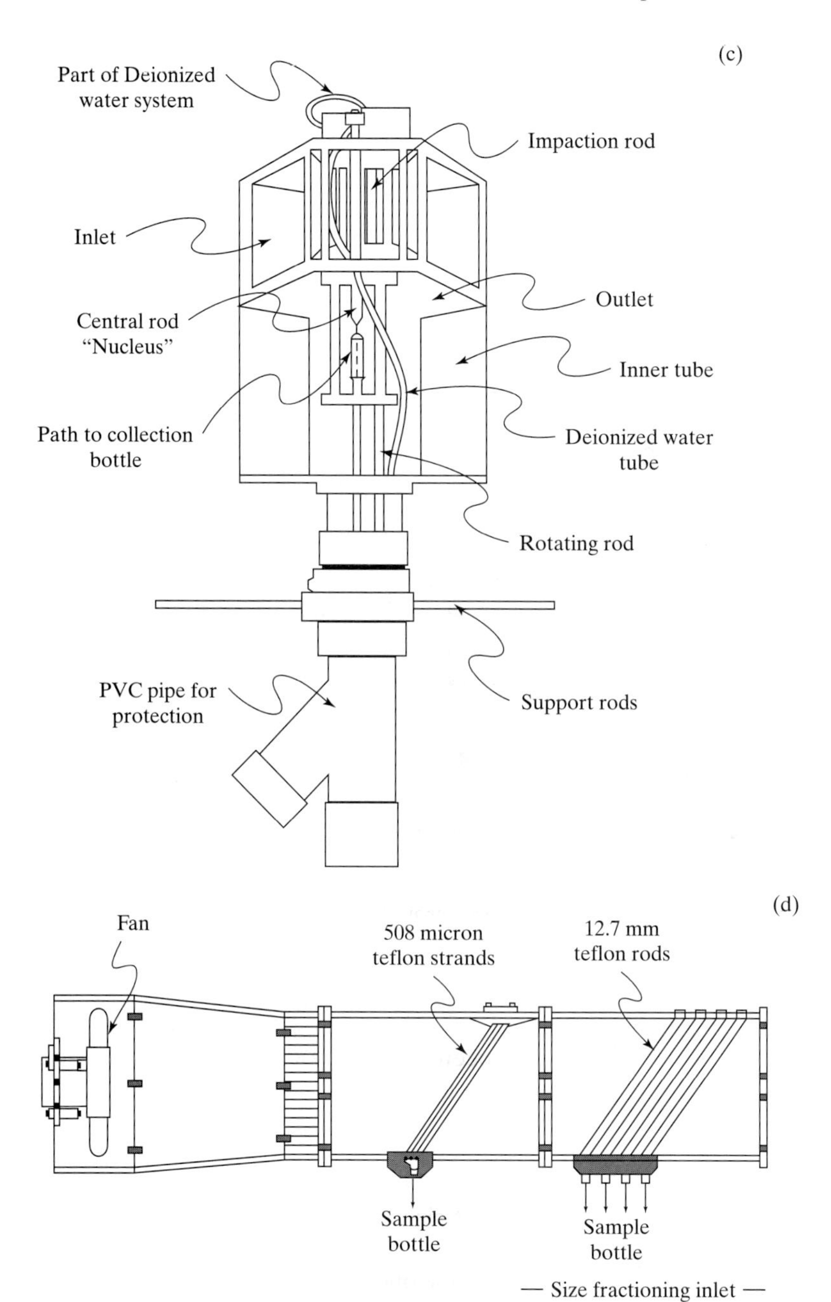


**Fig 7.4** *continued*

simultaneously collects small (≈9 $\mu$m) and large (≈23 $\mu$m) size cloudwater droplets (Demoz et al., 1996), and can operate in winds below 3 m $s^{-1}$. In contrast the DRI impactor collects only very small to medium sized cloud droplets.

Automated collector prototypes that collect samples every ≈20 min. are presently being developed. A prototype system with nearly continuous cloudwater chemical analysis is used at the Atmospheric Sciences Research Center, Whiteface Mt. Weather Observatory.

***7.1.2. Electronic.*** The electronic techniques will either yield the components of Equation 7.2 which are then used to estimate LWC, or they will yield a more direct measure of the LWC. A device known as a forward scattering spectrometer probe, (FSSP), (Fig. 7.5), or a 2-D variant of the FSSP known as an optical array probe, (OAP), (Fig. 7.6), photoelectronically determines the hydrometeor size distribution of the clouds. Both devices are manufactured by Particle Measurement Systems, Boulder, Colorado, a member of Fairey Group plc, Egham, Surrey UK. The hydrometeor size distribution is ideal because it includes information about the number densities of each droplet size category in the cloud, and this data can then be inputted into Equation 7.2b to calculate the LWC. The Johnson–Williams probe (Fig. 7.7), the King hot-wire probe (Fig. 7.8), and the Rosemount icing detector (Fig. 7.9) output a value of the LWC. The OAP has an added advantage over the FSSP of being able to distinguish between the phases of the cloud hydrometeors, and its theory of operation will be summarized in the electronic subsection under precipitation later in this chapter.

The FSSP and its operation are described in great detail by Baumgardner (1983), and Dye and Baumgardner (1984). Each FSSP may be configured for ground or more naturally airborne platform measurements. The ground-based version comes equipped with an inlet horn, sample tube accelerator, and honey comb flow straightener. The aircraft version does not require the horn and its optical

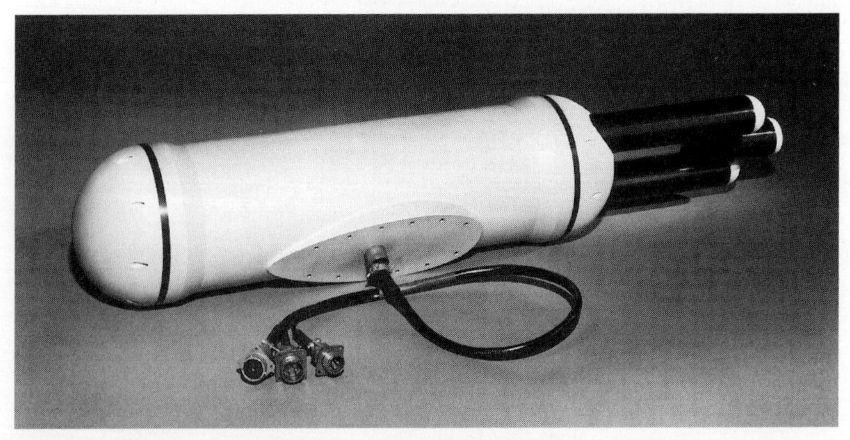

**Fig. 7.5** *A photo of a forward scattering spectrometer probe, FSSP. (Courtesy of Particle Measuring Systems, Inc.)*

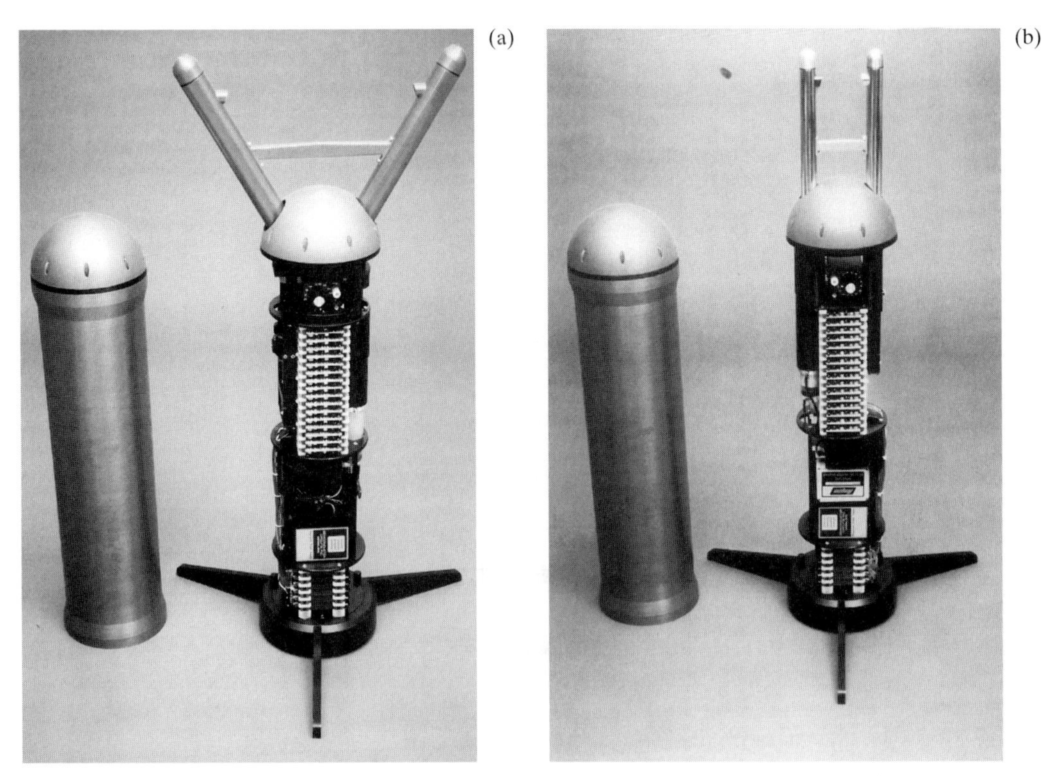

**Fig. 7.6** *Optical array probes for (a) precipitation, and (b) cloud drops. (Courtesy of Particle Measuring Systems, Inc.)*

heaters are configured differently. The FSSP measures the frequency distribution of cloud droplet sizes—i.e., the diameter and number of droplets that were detected of that size. The most frequent kinds of errors include: (1) coincidence (two particles counted as one), (2) deadtime (particles not counted while the probe is processing data), (3) wind gust, (4) reduced cutoff size due to low flow rate, and (5) probe orientation error. These errors result in (1) oversizing and undercounting, (2) undercounting, (3) overcounting, (4) biasing toward smaller sizes, and (5) biasing towards small sizes of droplets and undercounting, respectively. The first two depend on the droplet concentration. The probe must be continuously pointed into the wind ($<\pm 10°$) so the error due to (5) is negligible. The FSSP data used must be dynamically corrected, because the transit time of droplets through the probe depends on the droplet concentration and the windspeed (Choularton et al., 1986). When the droplet passes through the center of the laser beam, it passes through the widest section and produces the longest transit time. Higher droplet velocities due to an increase in sample flow decrease the transit time of the droplet. The percentage of time during which the FSSP is busy analyzing particles in the laser beam, or the *activity ratio*, equals the "actual number of valid counts" divided by the "total strobe count." The "total strobe count" indicates every droplet within the entire

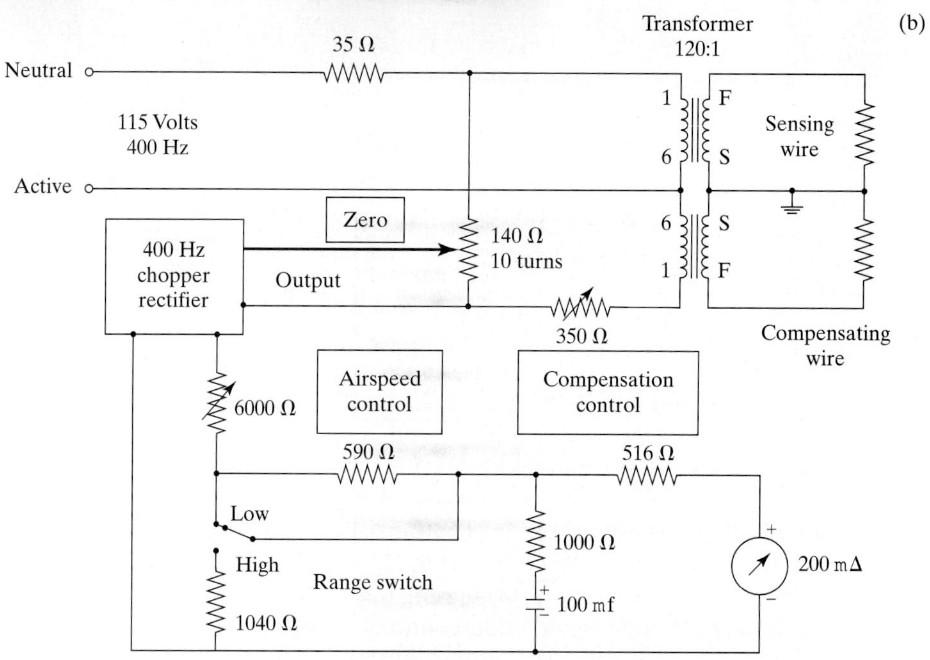


**Fig. 7.7** *(a) Hollow tube at the end of the lower extended arm is a Johnson–Williams hot-wire probe (Photo courtesy of Tom Henderson, Atmospherics, Incorporated); (b) a circuit diagram of that device.*

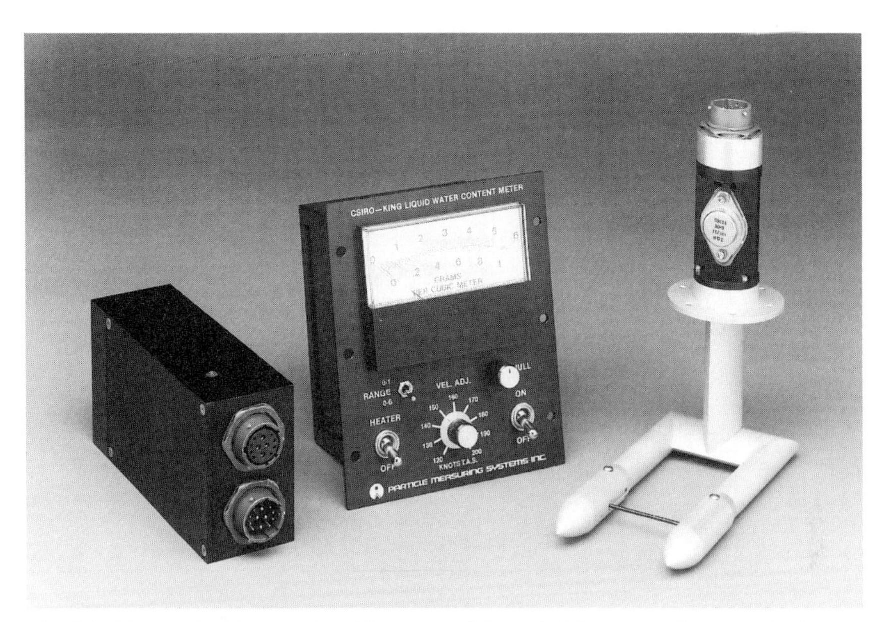


**Fig. 7.8** *Photo of a King probe. (Courtesy of Particle Measuring Systems, Inc.)*

(a)

**Fig. 7.9** *Photo (a) and schematic (b) of a Rosemount icing probe. (Courtesy of Tom Henderson, Atmospherics, Incorporated.)*

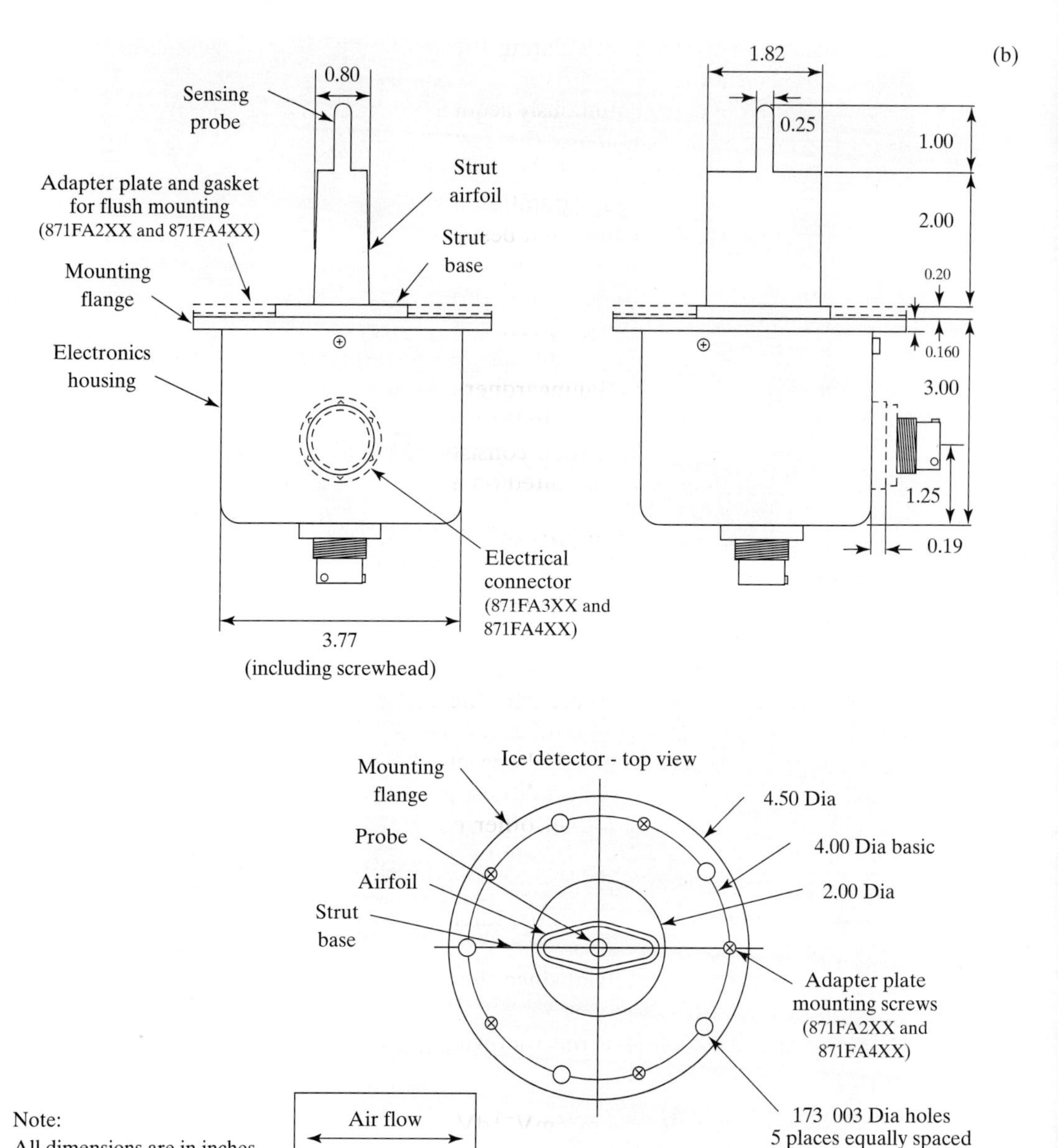


**Fig 7.9** *continued*

depth of field. The dynamic correction consequently involves multiplying the activity ratio by the total sample area to yield an effective sample area. The activity increases as droplet concentration increases causing coincidence losses of actual counts on an increasing percentage basis with activity. The effective sample area multiplied by the ventilation speed (typically $\approx$25.2 m s$^{-1}$) gives the volume sampled. The volume sampled multiplied by the concurrent activity ratio yields the actual volume used when calculating the microphysical parameters, such as the droplet concentration.

The FSSP data are continuously acquired every second by a software-controlled pulse height analyzer/accumulator that may be configured to send 1 to 999 s totals to disk. A word of caution, accumulating 1 s data requires megabytes of storage memory depending on the sampling duration. The software is nice because it allows the user to select the sampling time that best fits the problem at hand. There have been some recent improvements to the FSSP which include eliminating the deadtime error, upgrading the electronics, and improving the identification of the hydrometeors in the sample volume (e.g. SpecInc, Boulder, Co.). Processing delays (i.e., primarily the deadtime error) limit the use of the FSSP in regions of droplet concentrations $>500\ cm^{-3}$ (Baumgardner, 1983). A SpecInc Device yields an optical array that is many times superior to the OAP by particle measuring systems.

The Rosemount icing probe consists of a sensing cylinder with a diameter 0.64 cm and length 2.54 cm mounted on an oscillating plate that is exposed to the airstream. The droplets that come into contact with the sensing cylinder freeze on it, causing the frequency of vibration to decrease by an amount proportional to the amount of mass increase. The decrease in vibration frequency yields a proportional increase in voltage. The voltage is then digitized to 1 mV precision and recorded. When a preset voltage threshold of about 5 mV is reached, the probe tip is heated for a fixed short duration to remove the ice, and the cycle is repeated. Data taken after the de-icing are not reliable until the probe's temperature has stabilized. It is also particularly worthwhile knowing that calibrations of LWC can differ from one Rosemount probe to another. Voltage output increases slightly with decreasing temperature during non-icing conditions. Nonetheless, the Rosemount probe (i) is considerably less expensive than other electronic LWC devices, (ii) is simple to mount and operate, (iii) is not likely to respond to small, dry ice particles, (iv) is potentially a more reliable measuring device at much lower LWC than other instruments, (v) has a response more linearly proportional to LWC, and (vi) works best in clouds with temperatures between 0°C and −40 °C. The "non-response" to small ice particles is beneficial since these ice particles cause errors in the LWC determination from other instruments such as the FSSP. Heymsfield and Miloshevich (1989) provide the following expression for determining the LWC from the Rosemount probe

$$LWC_{rosemount} = [0.06\ g\ m^{-2} mV^{-1} dV/dt (TAS\ E''')^{-1}] - [0.0019\ g\ m^{-3}] \quad (7.3)$$

where $dV/dt$ is the time rate of change of voltage with units of $mV\ s^{-1}$, TAS is the true aircraft speed with units of $m\ s^{-1}$, $E'''$ is a mass weighted collection efficiency obtained using concurrent FSSP data. Cripps and Abbott (1997) describe a capacitive ice-sensing probe which shows promise in providing a more inexpensive, reliable alternative to the aforementioned icing probe.

The principle of operation of the Johnson–Williams and the King probes is based on the maintenance of a heat balance at a heated sensing element upon which water droplets impinge. The Johnson–Williams probe senses changes in the electrical resistance of the heated element as it cools due to the evaporation of the water, whereas the King probe measures the amount of power necessary to maintain a constant heated element temperature. King et al. (1978) provide additional details of the King probe operation, and the reader is reminded of section 5.1.2. Both of these hot-wire probes are sensitive down to $\approx 0.02$ to $0.05$ g m$^{-3}$ (i.e., their detection threshold) below which noise inherent in the instrument and drift due to changing airspeed masks any signal (Heymsfield and Miloshevich, 1989). Heymsfield and Miloshevich (1989) provide additional details on both of these hot-wire probes.

Baumgardner (1983) found the FSSP to yield higher LWC values than a version of the King and Johnson–Williams probes, and an impaction method, primarily because the FSSP concentration and size measurements were $\approx 17\%$ accurate. The hot-wire devices had accuracies that ranged from 5 to 20% compared with 30 to 40% accuracies with the FSSP and the impaction method. The Johnson–Williams method had an accuracy that was estimated to be better than 20% in a wind tunnel, otherwise it ranged from 20 to 40%. Saxena et al. (1989) found the integrated liquid water content from FSSP to agree within $\approx 7\%$ of the empirical liquid water content determined from the volume of water collected by the ASRC (DeFelice and Saxena, 1990). An empirical LWC relationship obtained from the ASRC collected water volume may be given by

$$\mathrm{LWC} = \{\mathrm{m}\,[(\text{collected water volume})\,(\text{windspeed time})^{-1}]\} + \mathrm{b}, \tag{7.4}$$

where m $= 17.374$ m$^{-2}$, collected water volume has units of g, windspeed has units of m s$^{-1}$, time is the duration of the sample, and b $= 0.059$ g m$^{-3}$.

Heymsfield and Miloshevich (1989) evaluated a number of liquid water measuring devices including the FSSP, Johnson–Williams and King hot-wire probes, as well as the Rosemount icing detector. They found that the LWC measurements from the icing detector and the FSSP compared favorably with those from the hot-wire probes in the range where the LWC is greater than 0.02 g m$^{-3}$. The hot-wire probes have detection thresholds that are about an order of magnitude higher than those possible with the FSSP and the Rosemount devices. The detection limit of the Rosemount probe is 0.002 g m$^{-3}$ (Heymsfield and Miloshevich, 1989).

Some comments with respect to spatial or temporal resolution are appropriate here. The samples collected while using the ASRC or the Caltech active string collectors are usually 0.5 to 1 h long, but could be as short as 0.1 h during a high liquid water content cloud. In comparison, the FSSP, Rosemount icing probe, and Johnson–Williams and King probes can yield LWC values every second allowing for more points to represent a particular time average.

The FSSP, or the SpecInc Device, would be applicable for the study outlined in Appendix E, along with a similar device to measure the aerosol spectra (e.g., diameters between 0.05 and $\approx 5$ $\mu$m). The Rosemount is not applicable, and the hot-wire probes do not provide the required drop size data. The cloudwater samples in Appendix E, particularly the final event sample, could be obtained with the ASRC collector as long as the windspeeds were $\geqslant\approx 2.3$ m s$^{-1}$. However, an active string collector such as the Caltech active strand size-fractionating cloudwater collector might be more appropriate.

An application of cloudwater measurements is the determination of the quantity of acids deposited onto a surface by clouds. This is generally accomplished by immediately preparing a cloudwater sample for a pH measurement, and then promptly splitting it into two (2) ionically clean sample bottles (typically 60 ml each) and refrigerated at 4 °C for later ionic analysis via ion chromatography consistent with U.S. EPA sampling protocols. The cloudwater deposition flux is estimated from,

$$F_{w,c} = I_c C_c, \tag{7.5}$$

where $I_c$ is the rate of cloudwater deposition obtained from a cloud deposition model (CDM) made to estimate the deposition onto a particular surface, and $C_c$ is the concentration of the desired species in the cloudwater sample. $I_c$ is very difficult to obtain accurately because it depends on the knowledge of the exact surface area upon which the cloud deposits. Cloud chemistry/deposition models are plentiful (Lovett, 1984; Tremblay and Leighton, 1986; NAPAP, 1987; Pandis and Seinfeld, 1989; Saxena and Lin, 1990; Wurzler et al., 1994; and many others). A CDM generally consists of a:

1. structural submodel that simulates the the morphology of the deposition surface;
2. hydrological submodel that computes cloud characteristics (i.e., microstructure, liquid water content) and dynamics, evaporation rate, and the cloudwater deposition rate resulting from cloud droplet impaction and sedimentation; and some also contain a
3. cloud chemistry submodel.

The cloud deposition rate has been estimated using eddy correlation techniques with very promising results (e.g. Beswick et al. 1991; Vong and Kowalski, 1994).

***7.1.3. Laser.*** A project entitled the experimental cloud lidar pilot study (ECLIPS) was initiated in 1988 to obtain information on cloudbase height, extinction, optical depth, fractional cloud cover, and surface fluxes (Platt et al., 1994). The measurements were made with visible, $CO_2$ and Raman lidars; a millimeter radar with a wavelength of 8 mm; a narrow field radiometer at the visible, infrared, and microwave wavelengths; standard wide field shortwave and longwave radiation flux instruments at the surface; a video system; and a rawinsonde. A lidar has also been used to identify subvisual-thin cirrus clouds for satellite research and climatological research (e.g., Sassen and Cho, 1992). The lidar data provide information on cloud base height, cloud depolarization ratio, and the aerosol profile (e.g., Spinhirne et al., 1996b; Chudamani et al., 1996). The depolarization ratio can be used to determine any differences in the microstructure (i.e., the number density of hydrometers of diameter D for all hydrometeor diameters) of clouds or to determine whether the clouds are precipitating, for example.

Lidar measurements have even been made at Palmer Station, Antarctica by Smiley et al. (1980) and at the South Pole (Smiley et al., 1976; Smiley et al., 1979; Smiley and Morley, 1979). Smiley et al. (1980) measured a depolarization ratio of 0.2 at Palmer on January 9, 1979 at 2300Z. Linear depolarization ratios for backscatter from various types of ice particles, 0.03 to 0.5 for mixed phase clouds, 0.5 for pure ice crystal layers, 0.5 for snowflakes, and 0.6 for rimed ice particles (Sassen, 1978). The

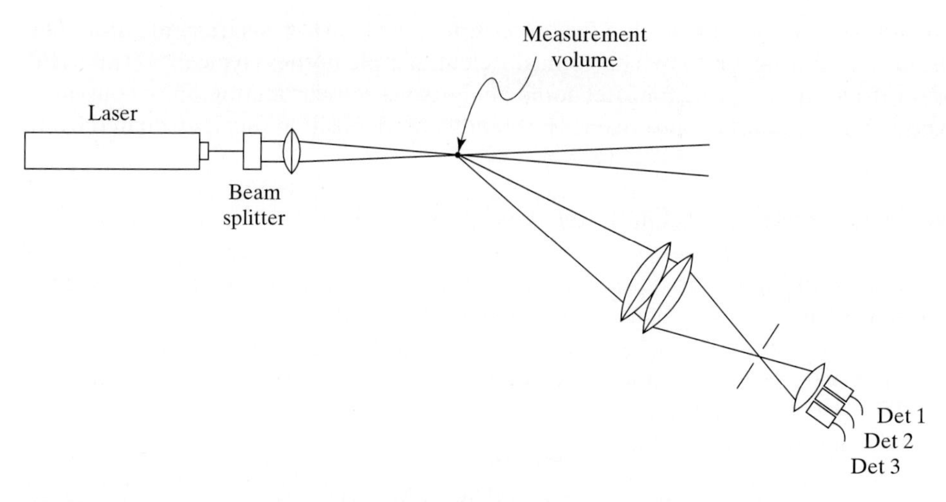


**Fig. 7.10** *Schematic of the optical component for a laser Doppler velocimeter and the droplet sizing system. DET denotes detector. (© Aerometrics, Inc. 1984. All rights reserved. Reprinted with permission.)*

depolarization for pure spherical water drop clouds should be zero for single scattering. Multiple scattering causes depolarization in dense water clouds. However, a high degree of orientation of ice crystals can produce very small values of depolarization (i.e., $<0.5$) that could be misinterpreted as water being present when there is actually none (Smiley et al., 1980).

Bachalo and Houser (1984) of Aerometrics Inc., describe a phase, doppler spray analyzer for simultaneous measurements of drop size and velocity distributions (Fig. 7.10). It is based on the measurement of the interference pattern produced by spheres passing through the intersection of two laser beams. The instrument has a linear response to drop size, and is independent of beam intensity and extinction resulting from surrounding droplets. The results are quite promising, and the instrument is experimental and very expensive.

***7.1.4. Satellite.*** The Global Hydrology and Climate Center, Marshall Space Flight Center scientists are conducting research on satellite sensing of precipitation, aerosols, and clouds. The cloud-related research includes the continued development of an algorithm to determine cloud-free pixels from those with clouds using Geosynchronous Operational Environmental Satellite (GOES) visible and two near infrared channels (Jedlovec, Guillory and Wohlman, personal communication, 1995). Ellrod and Nelson (1996) have used GOES visible (channel 1, 1 km resolution), infrared (channel 2, 4 km resolution) centered at 3.9 $\mu$m, and the window infrared (channel 4, 4 km resolution) centered at 10.7 $\mu$m to detect regions of potential aircraft icing in stratiform clouds with some promising results. Stearns et al. (1995) uses IR images to study cloud systems in the Antarctic region.

A satellite-borne sensor called the cloud and earth's radiant energy system instrument, or CERES, has been designed to yield the following output products:

cloud properties (i.e., amount, height, thickness, shortwave and longwave optical depth, cloud droplet or ice particle size and phase); top-of-atmosphere, atmosphere and surface radiative fluxes (both shortwave and longwave); and angular-dependence models of solar and thermal infrared radiation. The CERES top-of-atmosphere net flux and equal-area (1.25° Latitude/Longitude at the equator) data were averaged: (i) monthly over the globe; (ii) monthly over a region (at one standard deviation); and (iii) daily over a region (at one standard deviation). The data have total systematic errors of (i) 2, (ii) 3, and (iii) 9 W $m^{-2}$, whereas the science requirements for the same values are (i) 0.2–1, (ii) 2–5, and (iii) 5–10 W $m^{-2}$. The spatial sampling noise for regional monthly means is $\approx$1 W $m^{-2}$. Additional details of the error of the CERES sensor products are discussed by Wielicki et al. (1996). Hucek et al. (1996) found very good agreement between the CERES-I cloud amount data with scene-dependent time interpolation of 3 h and the GOES-derived 8 km narrowband radiance data.

Microwave satellite radiometric data have been used to determine the liquid water amount of clouds over the ocean (Grody, 1976; Staelin et al., 1976; Chang and Wilheit, 1979; Curry et al., 1990; Ferraro et al., 1996; and others) and land (Jones and Vonder Haar, 1990; Ferraro et al., 1996; and others). Ferraro et al. (1996) have also used microwave data to determine monthly cloud frequency over the globe. Curry et al. (1990) combine SSM/I microwave data with concurrent U.S. Air Force three-dimensional nephanalysis (3DNEPH) results to partially verify SSM/I-derived liquid water amounts (or paths), cloud type, cloud fraction, and cloud height. They show the cloud liquid water amount along the line of sight of the radiometer, or the liquid water path (LWP) may be expressed using the 18 GHz and 37 GHz channel brightness temperatures as:

$$\begin{aligned} \mathrm{LWP} &= \mathrm{LWP}_{37}, && \text{if } \mathrm{LWP}_{37} > \mathrm{LWP}_{18} \\ \mathrm{LWP} &= (0.33\,\mathrm{LWP}_{18}) + (0.67\,\mathrm{LWP}_{37}), && \text{if } \mathrm{LWP}_{37} < \mathrm{LWP}_{18} \text{ and } \mathrm{T}_{ct} > 263\,^{\circ}\mathrm{K} \\ \mathrm{LWP} &= (0.5\,\mathrm{LWP}_{18}) + (0.5\,\mathrm{LWP}_{37}), && \text{if } \mathrm{LWP}_{37} < \mathrm{LWP}_{18} \text{ and } \mathrm{T}_{ct} < 263\,^{\circ}\mathrm{K} \end{aligned} \tag{7.6}$$

where the subscript indicates the frequency channel used to derive the liquid water path (LWP), $T_{ct}$ is the cloud top temperature, and LWP is expressed in terms of cloud transmittance by

$$\mathrm{LWP} = \mathrm{A}^{-1} \ln(\mathrm{t}_c^{-1}) \cos\Theta, \tag{7.7}$$

where A is the absorption coefficient of liquid water and a function of frequency and temperature, and $t_c$ is the cloud transmittance.

One additional note on the retrieval of cloud information from remote sensing devices; especially from those devices on a satellite platform. The accuracy of the cloud properties retrieved from satellite measurements might be very comparable to the same variables derived from conventional devices. However, be aware that the satellite-derived quantities are not only determined from models that deconvolve the desired parameter, but those models and the satellite data itself are not necessarily "trained" (Kidder and Vonder Haar, 1995) with a representative dataset, or with a dataset of sufficient accuracy.

## 7.2. Precipitation

**7.2.1. Background** Precipitation measurements are made in order to obtain information about its temporal and spatial characteristics, such as phase, size distribution, intensity, duration, amount, and frequency (e.g., Hallett and Mossop, 1974; Pitter and Pruppacher, 1974; Kikuchi and Hogan, 1976; Czys, 1991; Czys and Peterson, 1992; Stewart, 1992; Mitchell, 1995). One could also add the additional reasons of acid deposition via precipitation (e.g., Hogan, 1982; Mayewski et al., 1987; National Acid Deposition Assessment Program, NAPAP, 1987; Lin and Saxena, 1991; Whitlow et al., 1992), the trace metal composition (e.g., Warburton, 1969; Warburton et al., 1980; and others) and isotopic content (e.g., Gedzelman and Lawrence, 1982; Warburton and DeFelice, 1986; Warburton et al., 1993) of the precipitation. Measurements of amount or depth are based on the assumption that the amount collected per unit area of the aperture of the gauge is the same as the amount collected per unit area that falls on the surrounding surface. Castelli began making precipitation measurements in Italy around 1639 (Salter, 1921), with an ≈12.7 m (5") diameter, ≈22.9 m (9") deep glass cylinder. Traditional precipitation measuring devices have changed very little since they were first made, except for their improved shape and accuracy. There are numerous references that provide details of the traditional precipitation gauges (e.g., BMOAM, 1961; Wang, 1975; Brock, 1984), and as a result, any comments pertaining to the traditional methods will be limited.

Precipitation is measured in terms of the total depth to which a flat, horizontal, impermeable surface of a known area would have been covered, assuming no water loss by melting, runoff, or evaporation. This depth is usually measured in inches in the U.S., but in millimeters (mm) in most other countries as well as in research applications. Frozen forms of precipitation (e.g., snow) are melted and added to any rainfall.

When snow is the observed form of precipitation, its depth is also measured. The British commonly use the approximation of 1' (or 1 foot) of snow yields 1" (or 1 inch) of liquid water, and the novice precipitation gauges obtained in the U.S. typically come with directions that 10" of snow equal 1" of rain. The latter water–snow conversion is generally known as the "water equivalent." The actual conversion depends on the temperature and moisture characteristics of the air, and the age of the snow once the snow is on the ground. Actually the conversion should be validated each time it is employed. The "snow density," $\rho_s$, is defined as the ratio of the water equivalent, WE, as measured with the gauge, to the depth of the snow, SH, as measured with a so-called snow stake. That is,

$$\rho_s = WE/SH. \tag{7.8}$$

Doesken and Jordan (1996) provide additional details on snow measurements. Chang et al. (1996) have used backscattered 35, 95, and 225 GHz polarimetric radar data from various types of snowcover to infer information about anisotropy, wetness, density, and average ice particle size of the snowpack. R. Stone, of the Atmospheric Sciences Center, Desert Research Institute, University of Nevada, Reno, still uses rectangular snow samplers to retrieve as many as one hundred 1 to 2 cm thick samples of fresh snow for trace silver (e.g., Warburton et al., 1980) and other chemical analyses.

Hail is treated differently than the other precipitation types (e.g., List, 1961; Wallace and Hobbs, 1977; Ludlum, 1980; Rogers and Yau, 1991; Young, 1993). Usually a number of additional characteristics are sought for hail that include: shape, roughness, structure, size, color (e.g., clear, opaque), density, surface temperature, and stable isotopic composition.

There are essentially two categories of precipitation instruments, namely, mechanical gauges (i.e., standard NWS rain gauge, a tipping bucket gauge, and the Belfort dual traverse weighing gauge) and remote sensing gauges. The mechanical gauges are the direct ancestors of Castelli's gauge with a number of accessories that include: wind shields, electric heaters for solid precipitation, and electro-mechanical counters for the tipping bucket gauges. The remote sensing gauges entail lasers, radars, satellites, and/or radiometers that are either ground-based or satellite-based, and possibly both.

The traditional standard NWS precipitation gauge has an 8" diameter opening for receiving the rain, that is converged into a 2.54" diameter, 20" inner tube with 0.01" or 0.2 mm graduated divisions. The cross-sectional area of the inner tube is one tenth of the area of the opening, resulting in a measured rainfall depth that is ten times that received by the collector. The amount of rain is measured by a ruler having graduations 1 mm apart (or 0.1" available). Each graduation is equivalent to 0.1 mm (or 0.01") of rainfall.

The dual traverse weighing gauge (such as that from Belfort, for example) measures the accumulated weight of precipitation in a bucket supported by a medium strength helical spring that is suspended from an adjustable arm. As the vertical column lowers, the pen moves upwards until it reaches the uppermost scale position on the chart (usually 6"). If the gauge continues to accumulate precipitation after it reaches the uppermost chart position, the mechanism controlling the pen location forces the pen back down the chart. This gauge is very accurate and reliable. However, the Belfort dual traverse gauge, especially, is very difficult to calibrate between 5" and 7" simply because this is the region where the pen changes direction, and any offset in this region is slightly amplified when the 12" amount is reached. This observation is based on personal experience and may be solely due to the age of the gauges being calibrated, but one should consider replacing some parts if a similar experience is encountered. The use of such gauges during periods of solid precipitation is ideally accomplished by placing a known volume of antifreeze solution (usually equivalent to $\approx$2" depth) into the collection bucket. The antifreeze also helps to inhibit evaporation. The use of a known volume of hot water to melt the solid precipitation is not operationally practical.

The tipping bucket gauge (Fig. 7.11) is very popular these days and consists of a collector, a triangular-shaped bucket, an electric transmitter, and a revolving drum. The bucket is balanced in bistable equilibrium about a horizontal axis and is divided into two equal compartments by a partition. It generally sits tilted to one side until there is a certain weight of rainfall caught (usually 0.01", but can be 0.1 mm). Once the preset weight of rainfall has accumulated, the bucket tips, empties, and exposes the other side to the rain. The tip takes 0.1 s. The bucket makes an electrical contact during the tip that can be recorded from any distance or sent to a datalogger. The tipping bucket gauge continuously measures the rainfall rate (flux of rainfall through a horizontal surface) with a $\pm$0.5% accuracy at

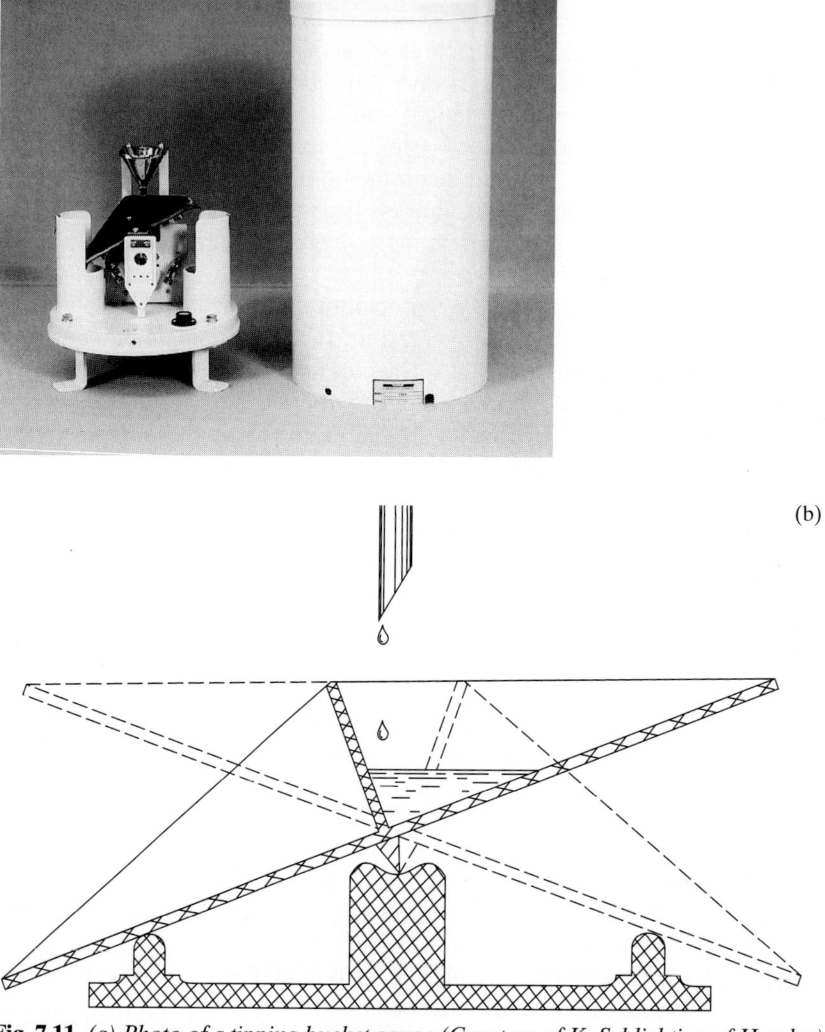


**Fig. 7.11** *(a) Photo of a tipping bucket gauge (Courtesy of K. Schlichting of Handar), and (b) closeup of the tipping bucket mechanism (Adapted from Ward, 1967).*

0.5 mm $h^{-1}$ to $\pm 3\%$ full scale between 25 mm $h^{-1}$ and 100 mm $h^{-1}$. The major disadvantages of the tipping bucket gauge include: (i) oscillation of the bucket when the windspeed is $\approx$15 mi $h^{-1}$ or higher, (ii) no record made for $<$0.01" accumulation, (iii) frequent calibration required, (iv) a high susceptibility to evaporation, and (v) splash errors during moderate to heavy rain showers. Most of the other rain gauges, especially those close in design to the tipping bucket, are prone to similar disadvantages.

How well does a gauge represent the amount of precipitation reaching the ground at a particular location? How well a gauge represents the precipitation amount reaching the ground depends on:

1. the density of the gauges;
2. the ratio of the diameter of the gauge to its depth (i.e., has been termed rim depth—bucket bottom to the rim);
3. whether or not the gauge has a wind shield;
4. the meteorological conditions during and throughout the time the gauge is in service; and
5. the placement of the gauge relative to its surroundings.

The accuracy or the amount of precipitation measured can be effected by evaporation, non-uniform diameter of the bucket, the graduated measure, and the placement of the gauge relative to its surroundings (i.e., its exposure is also important and may effect the representativeness of the reading). The amount of precipitation measured is inversely proportional to the gauge height primarily because of the relation of wind with height. The higher the gauge, the greater the windspeed and turbulence above the exposed surface and the smaller the catch. Wind shields of various designs are often provided to minimize the effects of wind on the collection of rain, snow, and hail. The distribution of precipitation corresponds to the topography and location of the continent in the macroscale analysis. Presently available precipitation networks are adequate for describing the seasonal variation of different climatic regions for most developed nations, but not the daily rainfall amount from a summer rain storm in a given region. The rainfall network with one gauge in an area of 500 to 600 $km^2$ (193 to 232 stat. $mi^2$), such as in the U.S., is not adequate. Note that 1 gauge per acre represents less than $8 \times 10^{-6}$ of an acre, and few areas have a density of gauges greater than one for every 6400 acres. In Great Britain there is one gauge for every 15 square miles on the average (Clements, 1964). The use of remote sensing techniques, such as radar with the gauge density appropriate to the local topographical features can improve sampling accuracy (Huff and Neill, 1957). Apparently, the laser gauge (section 7.2.3) does not have such problems, even though it does have the representativeness problem.

***7.2.2. Electronic.*** The electronic techniques are variants of the FSSP mentioned above for clouds. One of the more versatile variants is the optical array probe (OAP). Optical array probes are actually 2-D versions of the FSSP and will distinguish between solid and aqueous phase hydrometeors, as well as provide their respective sizes and number densities. Simply, the OAP (Fig. 7.6) uses a linear array of silicon photodiode elements as a sensor that is illuminated by a laser beam via suitable beam-forming optics and mirrors. The beam is directed outward from the probe enclosure through a hollow tube that has a mirror at its tip to reflect the beam across the sampling region of the probe to the tip of another tube that also has a mirror to reflect the beam back into the enclosure and onto the array sensor.

The array sensor is oriented normal to the flow of precipitation hydrometeors that travel across the beam between the probe tips. The passage of each hydrometeor shadows various array elements depending on its size, shape, and position during transit. The signals from the elements are processed in parallel to characterize each hydrometeor as a function of the decrease in signal detected from shadowed elements. The array elements are spaced 200 $\mu$m centers with 24, 32, or 64 elements for OAP models providing 15, 30, or 62 size channels, respectively. The actual sizing resolution in $\mu$m per array element (m/e) of a specific OAP is selectable and ranges from 200 to 10 m/e. The two dimensional versions not only report the size, but they also store the complete 2-D image as a series of 32-bit image slices during hydrometeor transit. The image and size data also allow one to determine the number density of the particular size hydrometeors. The OAP probes can be configured to sense cloud droplets, and they are expensive. The equivalent imaging probe by SpecInc is more accurate and about three times as expensive as the OAP probes.

***7.2.3. Laser.*** A lidar has been widely used to obtain characteristics of ice phase precipitation hydrometeors since Sassen (1978) and perhaps even earlier. However, it is only just within a few years (i.e., since ≈1994) that a reliable, low-cost, and accurate laser rain gauge is commercially available (Wang and Crosby, 1993). Handar carries optical precipitation gauges (Fig. 7.12) that are free of evaporation and splash errors; have a quick time constant; no moving parts; operable on a small amount of power, including a solar battery, and on land or water; and, most importantly, provides accurate estimates of instantaneous precipitation rate. These gauges measure rainfall rates between 0.5 mm $h^{-1}$ (i.e., mist) to 2,000 mm $h^{-1}$ (i.e., an extremely heavy, pushing, unnatural downpour) with ±5% accuracy between 1 to 500 mm $h^{-1}$ using only 25 mA (0.3 W). They have a 15 s time constant; are rugged, lightweight (5 lbs); have a dependable, self-operating design; and have a simple calibration check.

The company that manufacturers the optical precipitation gauges (Scientific Technology Inc.) also makes optical devices that measure snow, ice pellets, and hail. The optical gauges still have the representativeness problems since the spatial sample volume is of the order of ≈6–24" long.

***7.2.4 Radar.*** The precipitation rate, R, or intensity, may also be related to the radar reflectivity factor, Z (i.e., see Chapter 2). Some examples include Pani and Jurica (1989); Rosenfeld and Gagin, (1989); Rosenfeld and Woodley (1989); Fujiyoshi et al. (1990), and others.

For example, Pani and Jurica (1989) found the following Z-R relationship in the summer convective clouds over west Texas

$$Z = 260\,R^{1.02}. \tag{7.9a}$$

In contrast, recall that a Marshall–Palmer distribution of raindrops over the continent has the form

$$Z = 200\,R^{1.6}. \tag{2.19a}$$

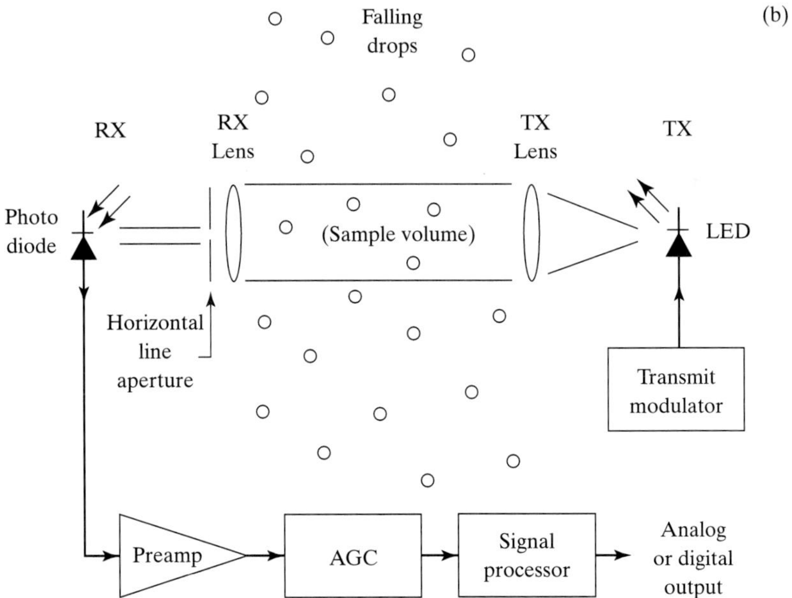


**Fig. 7.12** *(a) Photo of an optical rain gauge, and (b)a block diagram of its operation. (With permission of Scientific Technology, Inc.)*

Pani and Jurica also report that North Dakota convective clouds have a Z-R relationship of

$$Z = 155\,R^{1.88}. \tag{7.9b}$$

Fujioshi et al. (1990) found the following Z-R relationship for snow

$$Z = 427\,R^{1.09} \tag{7.9c}$$

where R is the snowfall rate and the radar did not have the same wavelength as that in Pani and Jurica (1989) for example. Recall that a Marshall–Palmer distribution of snow over the continent has the equivalent form

$$Z = 2000\,R^{2}, \tag{2.19b}$$

except R in Equation 2.19b was the rainfall rate of the equivalent melted diameters of the snow. One should know what kind of precipitation, where it will be observed, and the radar wavelength before applying and comparing Z-R relationships.

According to Ralph et al. (1995), spectral moment data with a 6 min. time resolution from National Oceanic and Atmospheric Administration 404 MHz radar wind profilers have been used to determine precipitation types (i.e., snow, light and moderate rain, convective rain, freezing rain), precipitation rates, and fall speeds. Their profiler results are quite promising and the system might be useable to validate SSM/I satellite precipitation data in the future.

***7.2.5. Satellite.*** Data from satellite-borne visible, infrared and microwave sensors have been used to determine precipitation amount, depth, frequency, and infer the kind of precipitation (snow, ice pellets, hail, rain, granpel) for over 30 years. Kidder and Vonder Haar (1995) provide a comprehensive accounting of the latter as they have done with all meteorological satellite observations. The use of Geostationary Operational Environmental Satellites (GOES) visible and infrared images to determine precipitation amount dates back to the early 1970s (e.g., Martin and Scherer, 1973), and extends down to the Antarctic continent (Stearns et al., 1995). Snowcover measurements have been made using satellite microwave brightness temperature data (Grody and Basist, 1996). Kummerow and Giglio (1994a, b) describe a passive microwave modelling technique for deconvolving rainfall amounts from space. The representativeness and accuracy of the precipitation measurements obtained via satellite-borne devices are like any of the other satellite-derived meteorological measurements—strongly dependent on the ability of the model to accurately deconvolve the appropriate data and the degree of the standard by which these measurements are validated.

When using the satellite data one must first determine at what wavelength the satellite sensor is operating, whether the satellite is over land or water, and, in this case, then determine whether there is precipitation or not. There are a number of different ways to proceed under each of these steps (see, for example, PIP-II, 1995). Usually, land surfaces and near land surfaces are flagged and verified using visible and IR threshold data values. Surface temperature and windspeed data used in a manner similar to Wentz (1991) will help filter the surface signal in the satellite's microwave brightness temperature.

Suppose our task is to generate precipitation rates over a portion of the globe in the mid-latitudes of the Northern Hemisphere using microwave sensors. The microwave satellite data can be obtained from data acquisition archive centers (DAACs) of NASA (e.g., Goddard and Marshall Space Flight Centers, see Kidder and Vonder Haar, 1995). The microwave data source is the defense meteorological satellite program special sensor microwave/imager (SSM/I). The SSM/I obtains radiometric data at 4 frequencies: 85.5, 37, 22.235, and 19.35 GHz for both vertical and horizontal polarizations, except 22.235 GHz having only a vertical polarization. The brightness temperatures obtained from the 19.35 GHz and 37 GHz (or $\lambda \approx 0.81$ cm) frequency channels are the primary source of information for quantifying oceanic rainfall rates of up to 25 mm $h^{-1}$ below the freezing level, especially when corrected for water vapor, wind roughness of ocean surface, and footprint averaging (Wentz and Spencer, 1996). The brightness temperatures derived from the 85.5 GHz (or $\lambda \approx 0.35$ cm) channel are a source of information for the ice phase (e.g., Spencer, personal communication, summer 1995; Mohr and Zipser, 1996). A surface tipping bucket rain gauge network dataset is available along with ground-based precipitation drop size spectra throughout the area covered by the gauges. The errors and problems involved with rain gauge data are discussed above, as well as by Petty (1997), Huffman et al. (1997), and others. The precipitation intensity from a satellite microwave sensor is commonly compared to, or trained with, the precipitation intensity determined from rain gauge data. For example, the $T_B$–R relationship involving two frequencies developed by Smith et al. (1992),

$$R = 125.5 - (0.435\,T_{B19}) + (0.108\,[T_{B19} - T_{B85}]), \tag{7.10}$$

for very strongly convective continental storms containing a cold-rain precipitation process was trained by rain gauge data. Equations 2.18 and 2.29a suggest that it might be more accurate to relate precipitation microphysical data (i.e., size, D, and number density of size D, phase) to the satellite data. Similarly, if liquid water path is related to brightness temperature (see Equations 2.21 and 7.2).

The microphysical measurements could be made using optical array probes or other relevant instrumentation adapted for the appropriate sampling platform (i.e., ground-based or airborne). The optical array data will also help quantify the relationship between brightness temperatures, ice and liquid water content, precipitation hydrometeor size and shape, and, of course, phase. For example, Smith et al. (1992) indicate that columnar-shaped graupel contents are best estimated at 15 or 19 GHz, and not 85 GHz. Note that graupel is commonly raspberry or conical-shaped. Evans and Vivekanandan (1990) have shown that shape has a significant influence on $T_B$'s at 85 GHz.

The primary problems encountered when retrieving precipitation characteristics over land using microwave sensors are the (1) lack of contrast between radiances from the strongly emitting land background and that from precipitation (especially rain) or a non-scattering atmosphere, and (2) strong spatial and temporal variability of emission from the land surface due to its irregular, thermally heterogeneous nature. These problems are generally handled by developing empirical screening criteria based on brightness temperature thresholds. For example, the 85.5 GHz horizontal brightness temperature threshold is at $\approx$238 °K and the

**Table 7.1.** *Conditions for classifying land pixels as non-raining*

| *Criteria for no rain* | *Description* |
|---|---|
| $T_{Bh85} > 265\,°K$ | No hydrometeors in the atmosphere |
| $T_{Bh85} > T_{Bh37}$ | No atmospheric scattering |
| $T_{Bv19} - T_{Bh19} > 15\,°K$ | Wet soil |
| $T_{Bh19} < 225\,°K$ and $T_{Bh85} > 150\,°K$ | Snow on ground |
| $T_{Bh85} < 230\,°K$ and $T_{Bh19} < 160\,°K = (0.4\,T_{Bh85})$ | Glacial ice and aged sea ice |

37 GHz vertical brightness temperature threshold is ≈240 °K if over land (Kummerow and Giglio, 1994a, b). An intense convective storm would have a 19.35 GHz brightness temperature of ≈250– ≈260 °K (Spencer et al., 1989). In contrast, Kidder and Vonder Haar (1977) determined that the brightness temperature at 19.35 GHz ranged between 125 °K and 175 °K over the ocean surface for a typical non-precipitating atmosphere.

The classification of land pixels as non-raining/raining may also be done in a number of ways. For example, Kummerow and Giglio (1994a, b) determine whether land pixels are raining or not if the criteria in Table 7.1 are met. Curry et al. (1990) found (a) liquid water paths of 350 and 500 g $m^{-2}$ were average thresholds for the onset of liquid precipitation in low and middle stratiform clouds, respectively, (b) their maximum rain rates were ≈7 mm $h^{-1}$, and (c) the auto-conversion of cloudwater to rainwater was determined to occur at a rate of 0.001 $s^{-1}$. Petty and Conner (1994) proposed an exploratory approach for retrieving precipitation characteristics over land using the SSM/I data. They look for and filter out the quasi-permanent and slow temporally varying components of the background emission. Briefly, they obtain 24 h averages of brightness temperature differences at a 0.25° latitude/0.33° longitude resolution in all seven SSM/I channels. These 24 h brightness temperature differences primarily reflect short-term variations in surface emissivity, temperature, and/or meteorological conditions; although some of the differences result from changes in snow consistency, soil moisture, and soil or vegetation temperature and precipitation.

Obvious precipitation events and snow cover are screened out based on $T_{Bv37} - T_{Bv85} > 5$ °K. Conner and Petty (1996) demonstrate that as long as 85 and 37 GHz vertically polarized channels (i.e., $T_{Bv37}$ and $T_{Bv85}$) are used then the remaining channels were relatively unimportant to the performance of their retrieval algorithm.

A criterion used to determine that a pixel, particularly over water, might contain "possible rain" is

$$P_{85} < 0.213 + \varepsilon_{85}, \tag{7.11}$$

where, $\varepsilon_{85}$ is an a-priori uncertainty assumed for all forward calculations of $P_{85}$. Petty (1994b) used 0.2 for $\varepsilon_{85}$, 0.1 for $\varepsilon_{37}$, and 0.05 for $\varepsilon_{19}$. P is defined in Petty and Katsaros (1990) and Petty (1994a, b) as the normalized polarization difference

$$P \equiv [(T_v - T_h)][(T_{v0} - T_{h0})]^{-1} \tag{7.12}$$

where $T_v$ and $T_h$ are the vertically and horizontally polarized brightness temperatures, and $T_{v0}$ and $T_{h0}$ are the vertically and horizontally polarized brightness temperatures for the same scene in the absence of all cloud. Additional details are found in Petty (1994b), including how $T_{v0}$ and $T_{h0}$ are determined. Petty (1994a) defined a scattering index that is related to P, and at 85 GHz it is

$$S_{85} = (PT_{v085}) + [(1 - P)T_c] - T_{v85} \tag{7.13}$$

where $T_c = 273\,°K$ and, P, defined by Equation 7.12, is the unpolarized limiting brightness temperature as a nonscattering cloud layer in the vicinity of the freezing level becomes optically thick. A similar expression to Equation 7.13 is obtained for the horizontally polarized channels by simply replacing the subscript v to h. A pixel could also be classified as "possible rain" if it is associated with $S_{85} > 10\,°K$. Values of $S \ll 0$ imply optically dense clouds that are predominantly warmer than 273 °K, or that the ice signature is *not* expected to be present. Values of $S \gg 0$ imply either large amounts of supercooled liquid water above the freezing level or scattering by ice. The maximum values of S in the latter case are unlikely to exceed 10 °K (Petty, 1994b). The scattering index at 85 GHz, $S_{85}$, is derived from the polarized corrected temperature (PCT) defined by Spencer et al. (1989), and it essentially responds to the brightness temperature depressions associated with volume scattering by precipitating ice hydrometeors.

The brightness temperatures derived from the 85.5 GHz (or $\lambda \approx 0.35$ cm) channel are a source of information for the ice phase (e.g., Spencer, personal communication, summer 1995; Mohr and Zipser, 1996). Large ice phase hydrometeors are poor absorbers but very good scatterers of microwave radiation (Spencer et al., 1989). As a result, the equivalent large ice phase hydrometeor precipitation rate is not retrieved from the satellite 85 GHz data, but from the PCT,

$$PCT = (aT_{bv}) + (bT_{bh}), \tag{7.14}$$

where $T_{bv}$ is the vertically polarized brightness temperature and $T_{bh}$ is the horizontally polarized brightness temperature. Given a frequency of 85.5 GHz and 53.1° earth incidence angle (typical of the microwave radiometer on the Defense Meteorological Satellite Program Satellite), then $a = 1.818$ and $b = -0.818$ (Spencer et al., 1989). The polarization correction removes the influence of the highly polarized, radiometrically cold sea surface from the relatively unpolarized scattering by precipitation-sized ice hydrometeors, and is normally applied to convective precipitating systems. Generally, low PCT accompany deeper ice layers, higher ice water contents, and increase as hydrometeor size increases. PCT frequency histograms are commonly used since the PCT has not been quantified in terms of ice abundances. The PCT is sensitive to temperature changes and there is a need to adjust retrievals based on temperature information (e.g., Spencer and Christy, 1992). Visual examination of 85.5 GHz imagery yields a threshold PCT near 266 °K that corresponds to the edges of deep convective cells. Thus the depression of the 85.5 GHz below the 266 °K threshold (Spencer et al., 1998), ΔPCT, namely

$$\Delta PCT = 266 - PCT, \tag{7.15}$$

might be used to detect the relative abundance of ice within the path viewed by the radiometer. A negative PCT depression indicates ice is more abundant than average, while a positive PCT depression indicates ice is less abundant than average. The PCT at 37 GHz (a and b are different by a few tenths of a unit or so compared to a and b at 85.5 GHz) is sensitive to graupel, whereas the PCT at 85 GHz is sensitive to dendrites (Spencer personal communication, summer 1997).

Other related problems encountered when retrieving precipitation characteristics using satellite microwave radiometer data include: (i) temporally and spatially mismatched scene areas from the various data platforms (i.e., satellite, airborne, ground-based measurements); (ii) precipitation retrieval over land surface that partially fills the scene of the satellite radiometric sensor, (iii) footprint-filling bias, that results when the precipitation partially fills the scene of the satellite sensor (e.g., see Petty, 1994a).

VIS and IR data have an equivalent swath width of ≈2,600 km. The scene or swath width of the SSM/I is ≈1,400 km with a resolution of 13 × 15 km at 85 GHz, to 43 × 69 km at 19.35 GHz. Mohr and Zipser (1996) defined mesoscale as approximately 10 SSM/I 85 GHz channel pixels, or about 2,000 $km^2$, and a mesoscale convective complex was defined as an area bounded by PCT = 250 °K of at least 2,000 $km^2$ of 85 GHz, with a minimum PCT ⩽ 225 °K.

Huffman et al. (1997) use a combination of ocean and land retrieval techniques to derive precipitation estimates for near coast pixels. The retrieval over the water body is generally accomplished using an emission retrieval technique (e.g., Spencer et al., 1989; Wilheit et al., 1991), while retrieval over the land is generally accomplished using a scattering retrieval technique (e.g., Ferraro et al., 1994). The emission technique eliminates land-contaminated pixels and so near coast estimates will generally result in relatively few pixels. Huffman et al. (1997) used the emission estimate if ⩾ 75% of the scene was water, otherwise they used a weighted interpolation between the two techniques to derive the precipitation rate, R, namely:

$$R_{combined} = [(n_{emiss} R_{emiss}) + (n_{scat} - n_{emiss}) R_{scat}] (n_{scat})^{-1} \quad (7.16)$$

where n is the number of samples, and "combined," "emiss," and "scat" denote combined, emission, and scattering estimates, respectively.

The beam-filling bias arises because of the inverse relationship between wavelength and antenna size required to achieve a desired ground resolution. Kidder and Vonder Haar (1985) state that footprints of at least 500 $km^2$ are too large to be filled by a uniform rain rate. Consequently, a brightness temperature averaged over a 500 $km^2$ footprint will generally be smaller than "true" since the brightness temperature varies directly with rain rate but in a non-linear manner. Wilheit et al. (1991) and Chiu et al. (1993) multiplied their rainfall rates by 1.5 in order to account for the beam-filling bias in monthly precipitation estimates.

The most recent development in the field of microwave radiometric sensing of precipitation is the advanced microwave precipitation radiometer (AMPR) which remotely senses passive microwave signatures from an airborne platform (Hood et al., 1994; Spencer et al., 1994). They describe AMPR (Fig. 7.13) as a low noise system with high spatial and temporal resolution at frequencies of 10.7, 19.35, 37.1, and 85.5 GHz. The ground spatial resolution is 0.6 km at a 20 km altitude for the 85.5 GHz channel, 1.5 km for 37.1 GHz, and 2.8 km for both 19.35 and 10.7 GHz

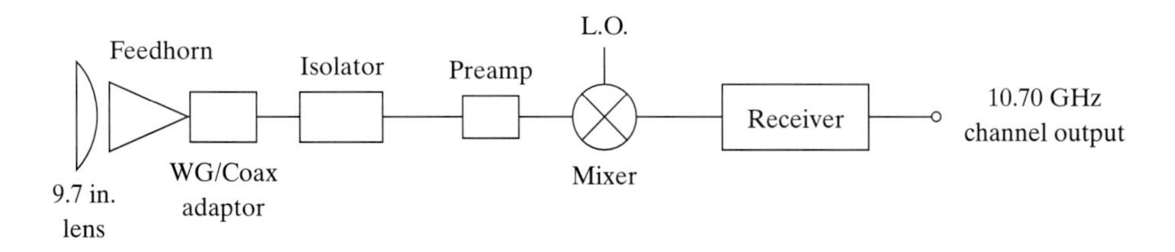


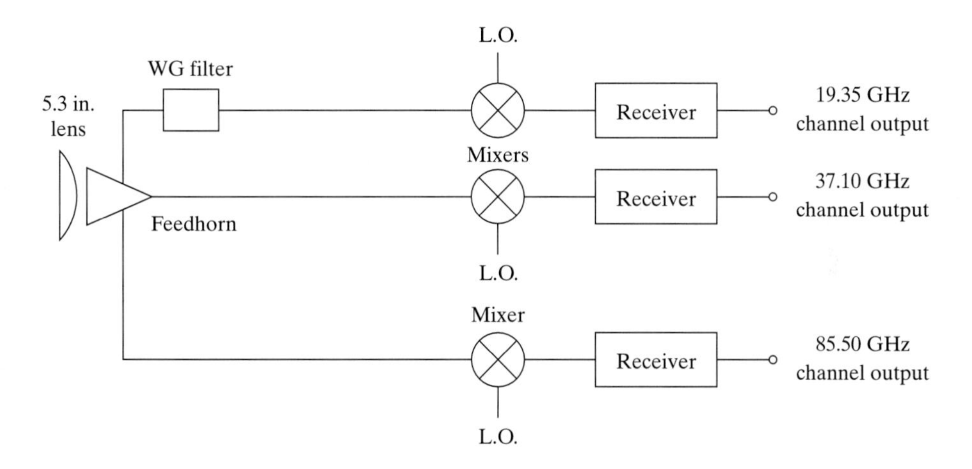


**Fig. 7.13** *A schematic of the receiving portion of the advanced microwave precipitation radiometer (AMPR) of NASA, MSFC. (After Spencer et al., 1994 with permission from American Meteorological Society.)*

channels. The AMPR has a 90° scan collecting a sample for each channel every 1.8° for a total of 50 samples per channel, yielding continuous footprints for the 85.5 GHz channel within a 40 km-wide swath, assuming an aircraft altitude of 20 km and an aircraft speed of 200 m $s^{-1}$. The AMPR frequencies are identical to the tropical rainfall-measuring mission, TRMM (Simpson et al., 1988). Snowcover measurements have been made using SSM/I data (Grody and Basist, 1996).

This chapter first introduced us to clouds and how to determine their water content and microphysical structure. The precipitation section then outlined the primary problems of using microwave radiometer data from satellite sensors to deconvolve information on precipitation characteristics.

The person using satellite data, whether to retrieve information about precipitation, clouds, aerosols, temperature, moisture, or atmospheric gases must not only understand how the individual measurements are made, their accuracy, representativeness, etc. They also need to know how this information may be combined with other independent, similar measurements to physically represent an entire satellite scene (Chapter 2), and not just an in-situ point as in most present day conventional devices. Once the latter is physically understood, one can then begin to use models to simulate the environments viewed by satellite and other measurement platform devices.

## Exercises

1. Calculate the radius of a droplet after 10 minutes of condensation growth in a steady-state environment with a supersaturation (saturation ratio in percent minus 100%) of 0.45%, P = 900 mb and T = 273 °K. Ignore curvature and solution effects.

2. Calculate the liquid water content of the following clouds.
   Cloud A : r = 10.1 $\mu$m and a corresponding N(r) = 50 $cm^{-3}$.
   Cloud B : r = 5.91 $\mu$m and a corresponding N(r) = 250 $cm^{-3}$.

3. Assume an ASRC and Rosemount probe simultaneously indicated a LWC = 0.216 g $m^{-3}$:
   a. What would the rate of water collection be for the ASRC (i.e., what is the collected water amount per time)?
   b. What is dV/dt for the Rosemount probe?
   c. Assume that the ASRC is also mounted on the same aircraft as the Rosemount and the windspeed past the ASRC is equivalent to the speed of the aircraft, then what would the rate of water collection be for the ASRC? (Assume that the ASRC is ideally set up so that none of the collected water freezes on impact and that there are insignificant loses due to evaporation, shedding, suspension of the droplets on the strings of the collector, etc.)

   The following information would be helpful in answering this question:
   TAS = 400 m $s^{-1}$, the efficiency of collection for the Rosemount probe is 0.5, and the windspeed = 10 m $s^{-1}$ at the ground where the ASRC sits in (a).

4. What is the deposition flux (g $cm^{-2}$ $h^{-1}$) of water and sulfate from a cloud with a liquid water content of 0.216 g $m^{-3}$ and a sulfate concentration of 6.25 $\mu$g $L^{-1}$, when the cloud-water deposition rate is 0.2 mm $h^{-1}$?

5. What is the water equivalent of snow if its:
   a. density is 0.8 g $cm^{-3}$ and depth is 10"?
   b. density is 0.95 g $cm^{-3}$ and depth is 10"?

6. Calculate the polarized corrected temperature derived from the following 85.5 GHz vertical and horizontal brightness temperature components.
   a. $T_{bv}$ = 270 °K, and $T_{bh}$ = 270 °K.
   b. $T_{bv}$ = 265 °K, and $T_{bh}$ = 270 °K.

CHAPTER EIGHT

# Complete Measurement Systems

This chapter is concerned with meteorological systems that contain one or more of the previously discussed measuring devices combined to fulfill a multi-purpose measurement function. Two examples of such systems are the rawinsonde and the automated weather system. There are two popular kinds of automated weather systems, namely, the automated surface observing system (ASOS) and the automated surface weather observing system (AWOS). The National Weather Service is in the midst of a modernization process that presently includes over 850 operational ASOS units. This chapter highlights the most significant operational features of the rawinsonde, AWOS, and ASOS units.

## *8.1. Sonde Measurements*

This section is primarily based on Lally (1985), Vaisala (1989), Garand et al. (1992), and Vaisala (1995a, 1997). The rawinsonde is simply a balloon-borne radiosonde that measures the windfield, and also consists of a transmitter, receiver, tracking unit, and a recorder. The radiosonde contains pressure, temperature, and humidity sensors and transmits air pressure, geopotential height, air temperature, and dewpoint depression back to the receiver where they are converted into appropriate units and sent to the recorder or storage device. Wind direction and windspeed are determined from the ascent rate of the balloon and tracking its position after release. Radiosondes come in a variety of designs and configurations, including those that are made to drop out of airplanes (i.e., dropsondes).

A typical U.S. radiosonde consists of a baroswitch; windmill; motor switch, or clock device that engages a temperature-compensated aneroid capsule to move a lever arm across a communicator plate; a lead-carbonate-coated rod thermistor about 0.7 mm in diameter and 1 to 2 cm long; a carbon humidity element that is similar to a dew cell hygrometer, except that the carbon replaces the LiCl (See Chapter 6); and a radio transmitter to relay its measured quantities to a ground-based radar, navigation aid, or other tracking system.

Advancements in electronic and navigation technologies have lead to recent improvements of the sensor and receiver portions in the typical radiosonde. The

**Table 8.1** *Measurement range, resolution, and accuracy of sensors on a recent rawinsonde*

| *Parameter* | *Range* | *Resolution/Accuracy* | *Response Time* |
|---|---|---|---|
| Pressure[a] | 1060 mb to 3 mb | 0.1 mb/0.5 mb | |
| Temperature[b] | +60 to −90 °C | 0.1 °C/≤0.2 °C | ≤2.5 s |
| Humidity[c] | 0 to 100% | 1%/±2% | 1.0 s |
| Windspeed | 0 to 180 m s$^{-1}$ | 0.1 m s$^{-1}$/0.3 m s$^{-1}$ | |
| Wind direction | 1 to 360 ° | 1 ° | |
| Receiver | GPS | 0.5 s | |
| | Loran-C | ≈120 s | |

[a]Capacitive aneroid or BAROCAP®.

[b]Capacitive bead or THERMOCAP®. Response time of 2.5 s under 6 m s$^{-1}$ flow at 1000 mb. See Table 3.5.

[c]Thin film capacitor or HUMICAP®.

range, resolution, and accuracy of the improved sensors are contained in the recent RS-80 radiosonde units manufactured by Vaisala and are shown in Table 8.1. The RS-80 temperature sensor is hermetically sealed to keep away moisture, and is electrically grounded with a thin aluminum coating deposited on the sensor capsule that will also minimize radiation errors (e.g., maximum of 2 °C at 1,000 mb and a 45 ° solar elevation) during observation. The RS-80 humidity sensor not only has a small thermal mass, but a very reliable and very fast response time as well. In fact, Vaisala is experimenting with an even faster temperature sensor for their RS–90 radiosondes (see Table 3.5). The three "cap" sensors (RS-80) and the RS–90 (Vaisala, Inc.) temperature and humidity sensors are shown in Fig. 8.1 and Fig. 8.2 respectively.

The latest receiver is the Global Positioning System (GPS). The Global Positioning System consists of 24 satellites orbitting at heights of ≈20,000 km that transmit radio signals at wavelengths of ≈19 and ≈22 cm through the earth's atmosphere to receivers on earth, as well as to low earth orbit satellites (Businger et al., 1996). The GPS receivers have become inexpensive and light enough to be carried on radiosondes. The ascending radiosonde receives the GPS signals, extracts the necessary information and transmits it over a narrow band width (Vaisala, 1995b). The telemetered information from the radiosonde is detected, amplified, converted by the receiver, and processed for storage or sent to a strip chart recorder or a computer-based data acquisition display. The tracking unit automatically locks onto and tracks the radiosonde unit. The elevation and azimuth angle information obtained by the tracker is used in conjunction with the height information from the radiosonde transmission data to compute winds aloft. The wind calculations combine the radiosonde derived GPS data and the locally received GPS navigational information (Vaisala, 1995b, 1997). A differential correction is applied to provide highly accurate (≈0.2–0.5 m s$^{-1}$) wind data with excellent vertical resolution (30–50 m), according to Vaisala (1995a). The development of the global positioning system for meteorological applications (e.g., Businger et al., 1996) has greatly improved the accuracy with which winds are determined from the rawinsonde system. The GPS system is used for the more research intensive applications. The GPS can provide very accurate position, time, and windspeed measurements, is not susceptible to atmospheric electrical interference, and allows for tracking of multiple

(a)

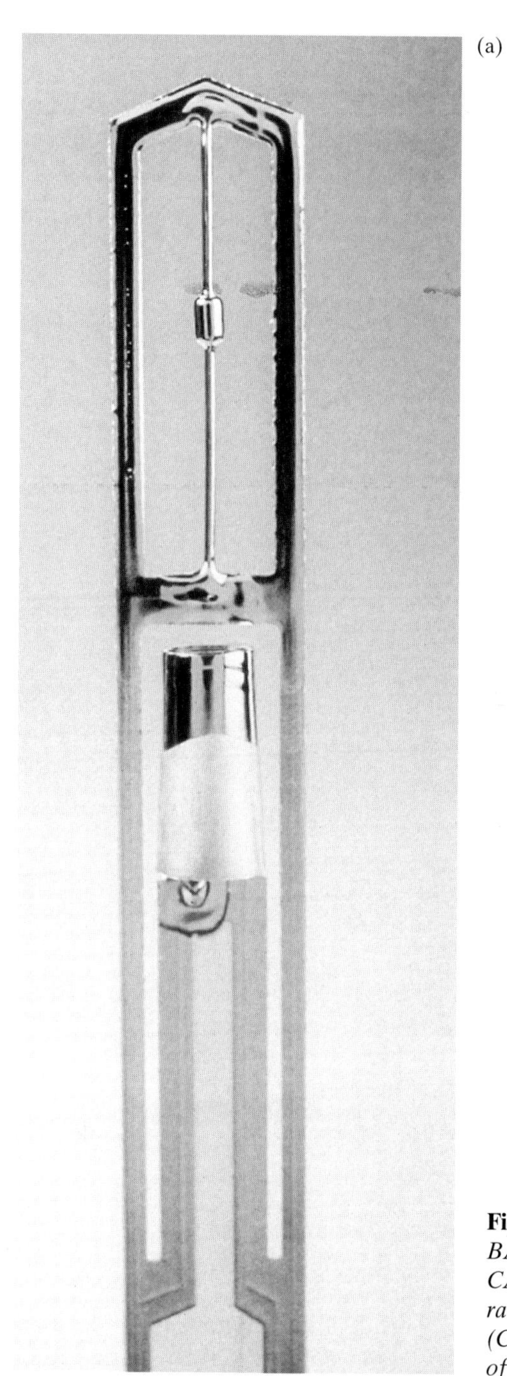

(b)

(c)

**Fig. 8.1** *The Vaisala BAROCAP®, THERMO-CAP®, and HUMICAP® radiosonde sensors. (Courtesy of K. M. Goss of Vaisala, Inc.)*

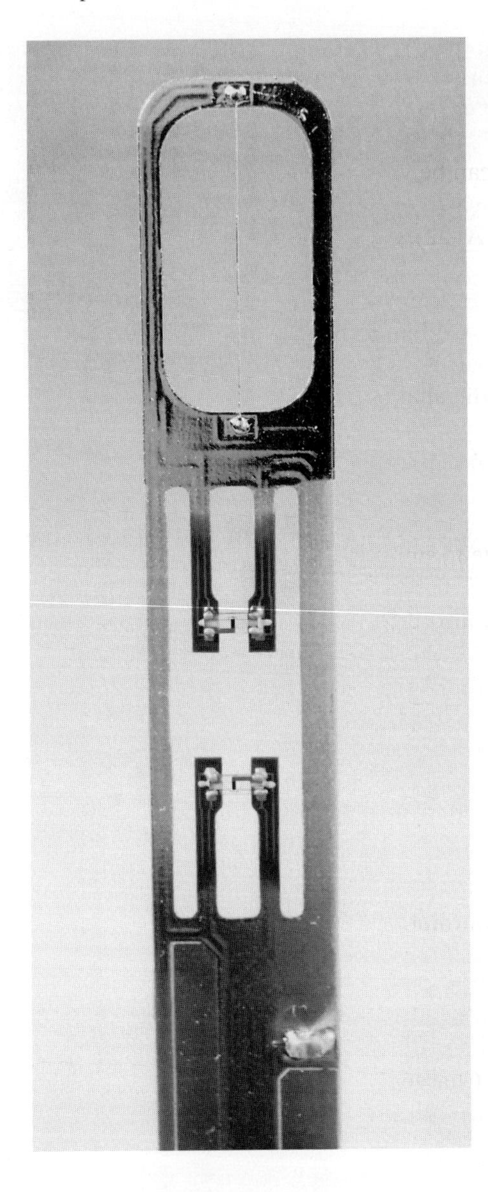

**Fig. 8.2** *The Vaisala RS-90 temperature and humidity sensors. (Courtesy of K. M. Goss of Vaisala, Inc.)*

balloons (Businger et al, 1996). However, the Loran-C navigation system accompanies a more affordable and nearly as accurate radiosonde system that can and has been used to teach undergraduate meteorology students (e.g., Colby, 1994).

Despite the very positive and long overdue improvements in sondes, their frequency of deployment is still far too small, and their spatial coverage still far too sparse and heterogeneous (e.g., the spacing between rawinsonde stations is closer in the northeast United States than in the western United States). The humidity sensor, which is designed to measure the amount of water vapor, still has difficulty working accurately in extremely cold temperatures due to ice formation.

Once the sonde reaches the point of neutral stability (i.e., the vertical atmospheric altitude at which it can no longer rise to higher altitudes), its balloon will ultimately burst and drop back to the earth. The sondes are equipped with parachutes that generally open as they fall. Furthermore, only 25% of the routinely released sondes are ever recovered, which can be very cost inefficient.

It is worth noting that improvements in technologies, different manufacturers (e.g., Vaisala, Metox, Graw, KAW, A–22), different calibration techniques, and even different reporting practices around the globe should make any operational, consultant, educational, research, or government meteorologist use global data archives with caution. We touched on this very subject at the end of Chapter 3 and subsequent chapters, especially those that contained data from remote sensing devices. Please, if you are involved, in any way, with the analysis of any dataset, especially one derived from different sources over the globe, make sure that your computations actually indicate the "true" changes in the data, and not the differences in the technologies and instrumentation, etc. If the latter is not possible then make it known. Eskridge et al. (1995) provide a very comprehensive description of an upper air dataset they assembled, and describe a technique to simultaneously detect vertical and horizontal errors in the meteorological variables derived from radiosonde and pibal data.

Remote sensing devices have also been used to obtain vertical moisture, temperature and windfield profiles. For example, Uttal et al. (1990) obtained very high resolution measurements of the water vapor fluxes using a scanning Doppler radar and a passive microwave radiometer. They point out three advantages of their vertical profile measurement system over the rawinsonde that are worth repeating, namely:

a. their radar / radiometer system provides instantaneous values of temperature, moisture, and of the windfield;
b. their remote sensing profile is a "true" vertical measurement; and
c. their system provides quasi-continuous vertical profiles of temperature, moisture and the windfield.

A system such as that of Uttal et al. is still experimental. The SPECTRE system of ground-based moisture, temperature, aerosol, and cloud profilers (Ellingson and Wiscombe, 1996) mentioned in Chapter 6 is another example. However, the rawinsonde data is and will be needed for a number of years to "train" the systems just mentioned.

## 8.2. *Automated Weather Systems*

Automated weather systems were created to provide a suite of measurements for mesoscale experimental programs, and they are becoming increasingly popular for placement in remote regions of the United States and elsewhere. They use one or more of the sensors described in Chapters 3–7. These systems are commercially available and come in a variety of configurations that will allow the user to meet almost any operational need, including those of hydrometeorologists. A typical remote environmental monitoring system might include a lightning rod; rain gauge; propeller anemometer that points into the wind; shielded relative humidity, shielded temperature, air pressure, and pyranometer sensors; a datalogger; and an internal battery or solar panel. Automatic weather stations (Stearns, 1996) and satellite data have been combined in the Antarctic region (Fig. 8.3) for operational forecasting and research applications (e.g., Stearns and Wendler, 1988; Stearns et al., 1995; Turner et al., 1996). Another exam-

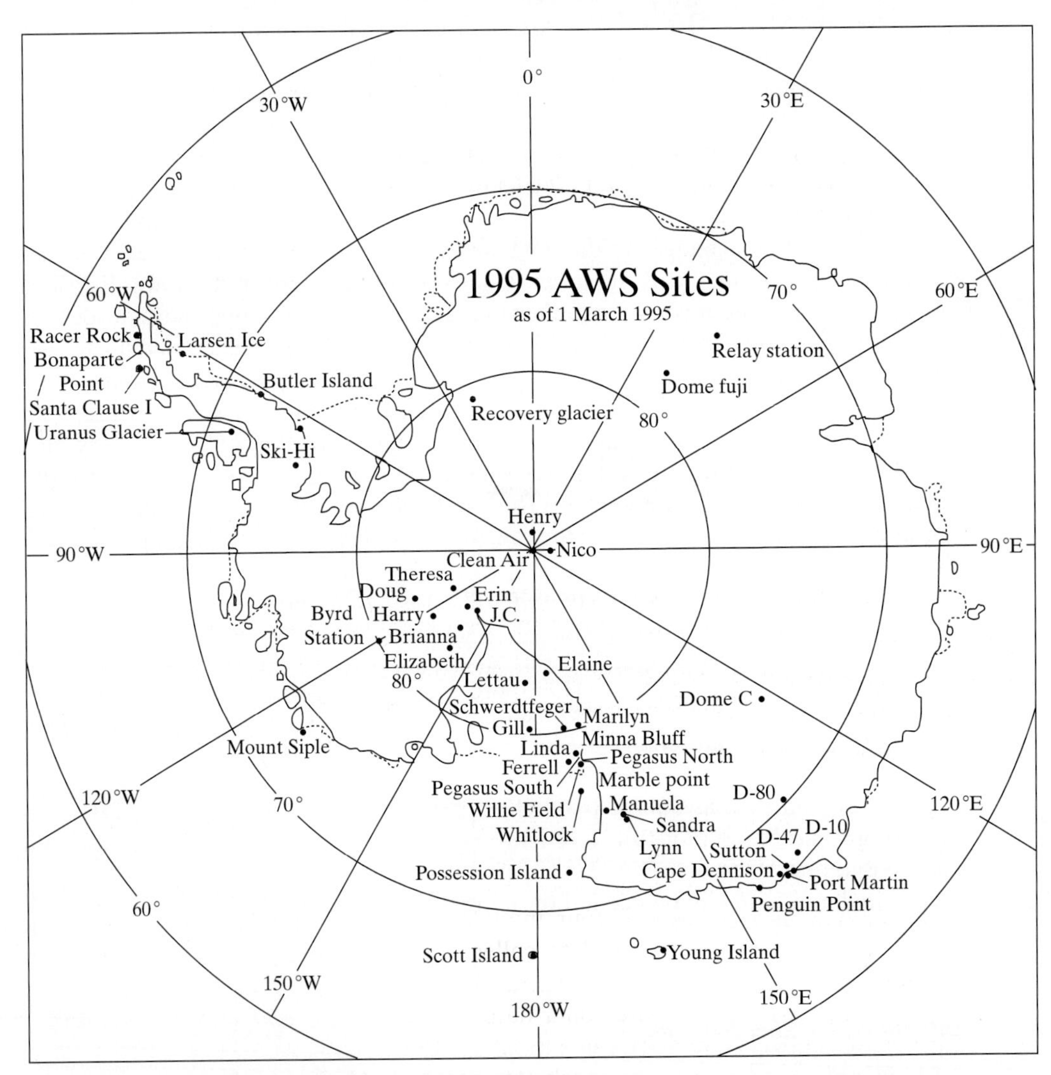


**Fig. 8.3** *The University of Wisconsin Antarctic automated weather stations for 1995. The stations located between 120° W through 180° through 150° E longitude and between the South Pole and about 76° S latitude make up a mesoscale network. The photo on page 155 shows the University of Wisconsin automated weather station at Doug site Antarctica, January 21, 1997. The Twin Otter aircraft provides transportation for installing and repairing the Antarctic automated weather stations. (Map and photo courtesy of Dr. C. R. Stearns, University of Wisconsin–Madison)*

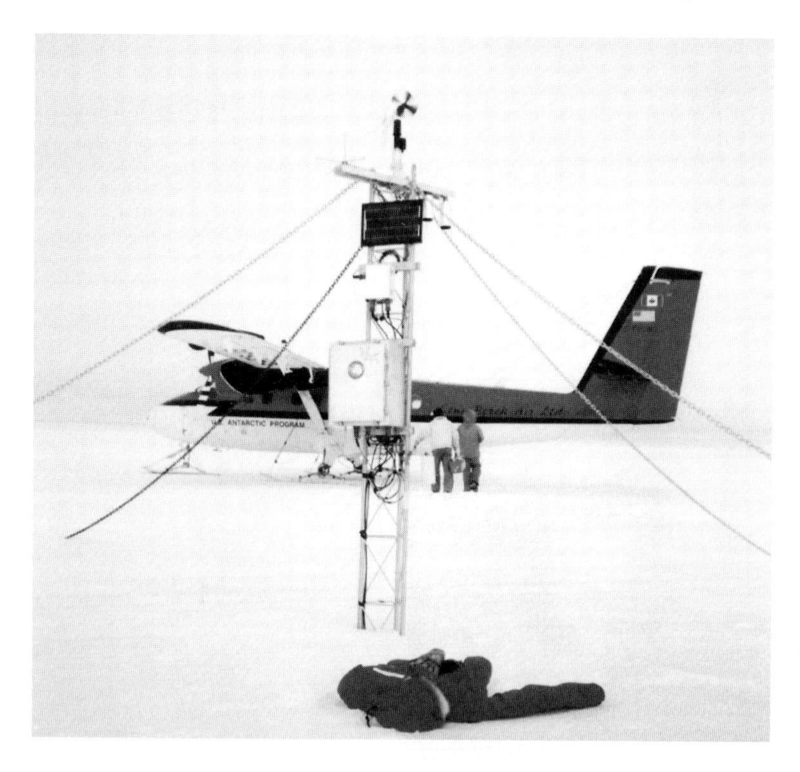


**Fig. 8.3** *continued*

ple of the multiple operational applications of automated weather systems lies in the differences between the automated surface weather observing (AWOS) system and the automated surface observing system (ASOS) used by the National Weather Service. Please note that Tanner (1990) provides an excellent overview of automated weather stations used in studying vegetation canopies, their standard instrumentation, and sensor performance.

***8.2.1. Automated Surface Weather Observing System.*** The automated surface weather observing system or AWOS (Fig. 8.4) generally provides real-time automatic weather updates, and may be deployed wherever weather data are scarce. The AWOS unit consists of four modules, a sensor module that measures the meteorological parameters, a data collection platform module that collects and preprocesses the sensor data, a central data platform module that completes the data processing and prepares data outputs, and a data distribution module that provides the final output data in a number of different forms (i.e., printer, ground-to-air-radio, telephone, operator's terminal, datalogger). The data collection platform is equipped with lightning protection and provides the interface between the sensors and a data processor. The processor queries each of the sensors once a second, conditions, and then converts the sensors inputs to digital format. Then it preprocesses the data by performing averages, checking parameter limits, and performing unit conversions. The data is then sent to the central data platform

**Fig. 8.4** *An example of an automated surface weather observing system (AWOS). The photo shows an AWOS III system at the runway touchdown location at Pipestone Municipal airport, Pipestone, Minnesota. The data from AWOS may be broadcast to pilots via a voiced discrete VHF radio frequency, or existing field NAVAIDs (i.e., VOR or NDB). The line drawing defines the various AWOS component systems. (Courtesy of Handar, Inc.)*

module via either a direct connection, modem, dedicated line, or a radio depending on the configuration of the AWOS and the needs of the user. The central data platform is typically where the data is applied to various algorithms depending on the application and then archived. The basic, or standard AWOS unit has the following sensors: air temperature, dewpoint temperature, pressure (altimeter setting), and windfield (i.e., windspeed, wind direction, and peak gusts). However, cloud cover, visibility, total radiation, and liquid precipitation amounts, for example, may be added if desired. The range and resolution of the typical AWOS system are shown in Table 8.2.

*8.2.2.* ***Automated Surface Observing System.*** The Automated Surface Observing System, or ASOS (Fig. 8.5), provides real time data and is generally configured into four simple modules as mentioned for the AWOS unit. The primary difference lies in the tasks that AWOS and ASOS were designed to address. For example, compare the parameters measured by a typical AWOS unit (Table 8.2) to those of a typical ASOS unit (Table 8.3). Now consider the other weather elements measured by ASOS: sky

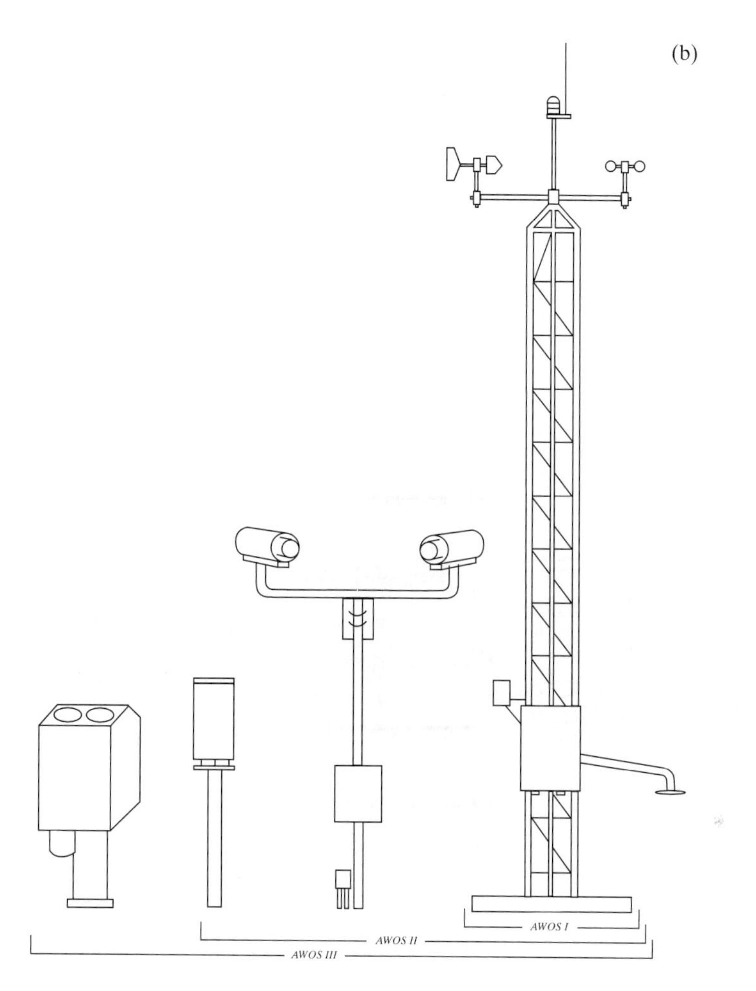


**Fig. 8.4** *continued*

**Table 8.2** *The range and accuracy of typical AWOS output parameters*

| *Parameter* | *Range* | *Accuracy* |
|---|---|---|
| Altimeter setting | 17.58" Hg to 32.48" Hg | 0.01" Hg |
| Temperature | −65 to +130 °F | 1 °F |
| Dewpoint | −30 to +90 °F | 1 °F |
| Peak wind gusts | 0 to 85 knots | 1 knot |
| Windspeed | 0 to 85 knots | 1 knot |
| Wind direction | 0 to 360° | 10° |
| Visibility | <0.25 to 10 miles | <0.25 to 3 miles[a] |
| Sky Condition (ceiling) | 100 to 12,000 ft. | 100 to 1,000 ft.[a] |

[a]Actually a resolution, and is commonly incremental.

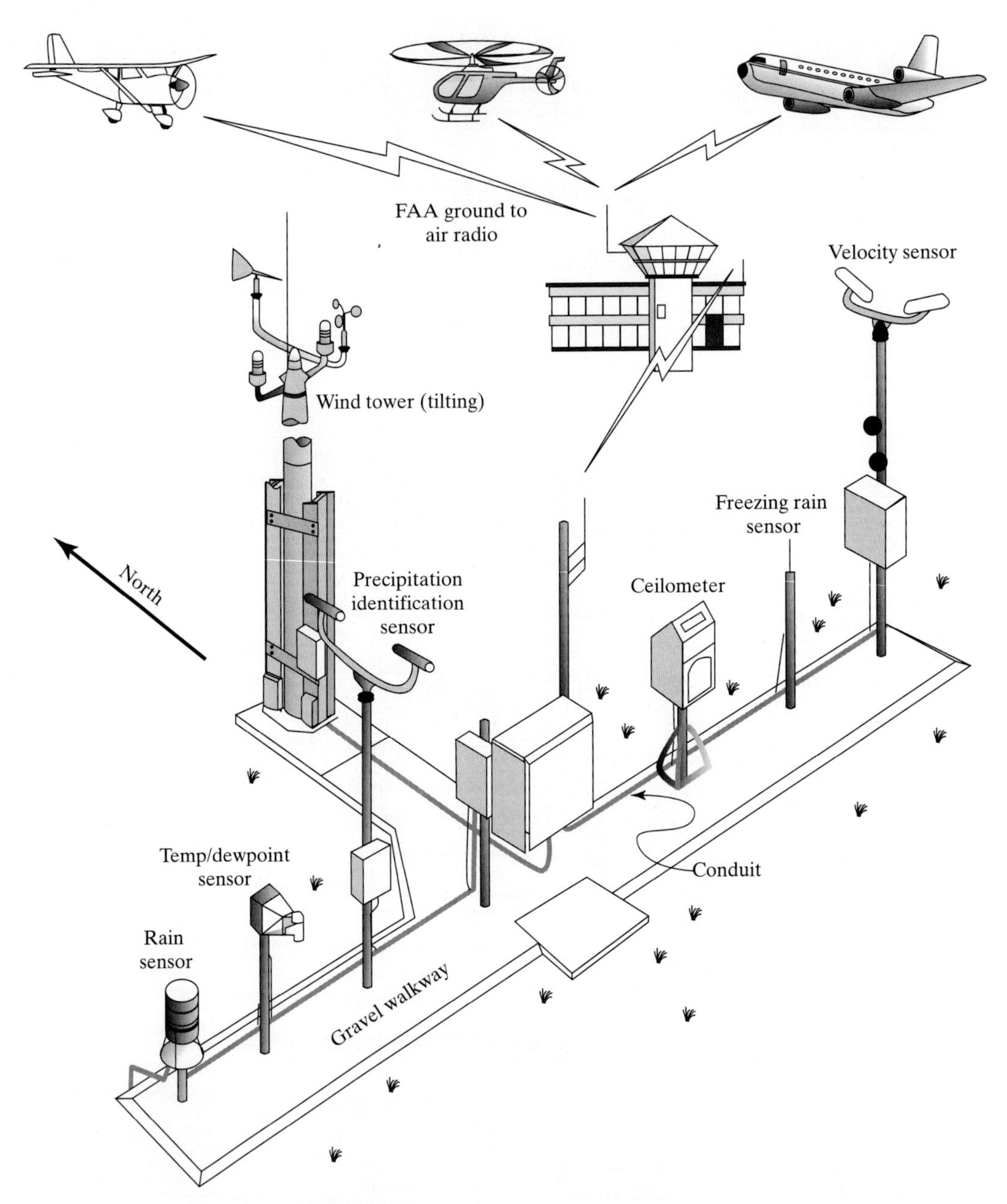


**Fig. 8.5** *An example of an automated surface observing system (ASOS). (Courtesy of R. Schaar, National Weather Service, Sullivan, Wisconsin.)*

**Table 8.3** *The range and accuracy of typical ASOS output parameters*

| *Parameter* | *Range* | *Accuracy* |
|---|---|---|
| Pressure | 16.9" Hg to 31.5" Hg | 0.1" Hg |
| Temperature | −65 to +130°F | 1°F |
| Dewpoint | −30 to +86°F | 1°F |
| Windspeed | 0 to 125 knots | 1 knot |
| Wind direction | 0 to 360° | 10° |
| Visibility (Forward scatter) | <0.25 to 10 miles | <0.25 to 3 miles[a] |
| Cloud height | 0 to 12,650 ft. | 100 ft.[a] |
| Precipitation | ⩾0.01 inches | 0.01 inches[b] |
| Freezing rain | ⩾0.01 inches | 99%[c] |

[a]Actually a resolution, and is commonly incremental.

[b]Actually a resolution.

[c]As long as there is at least 0.01" of ice accumulation.

condition (i.e., clear, scattered, broken, overcast), fog or haze presence, and significant remarks (i.e., variable cloud height, variable visibility, precipitation beginning/ending times, rapid pressure changes, pressure tendency, wind shift, peak wind—also mentioned in Table 8.2). The ASOS system is designed to replace, or take the place of, a National Weather Service (NWS) Observing Station. The ASOS information will:

1. help the NWS to improve the accuracy and timeliness of its forecasts and warnings;
2. improve airport safety since they are mostly placed in airport runway touch-down zones;
3. detect significant changes in weather parameters;
4. disseminate hourly and special observations via the NWS and Federal Aviation Administration communications networks; and
5. provide up-to-date weather information and altimeter setting reports directly to pilots during approach and landing, along with easier phone access to weather for preflight planning.

However, weather observation duties will ultimately be eliminated.

The ASOS unit is ideal for regions on the globe where data are unavailable (such as mountains and airports), and any reliable, continuous data are welcome. According to the ASOS operations and management center (e.g., Lojek, 1996), the ASOS unit has been plagued by a number of complaints, such as:

a. sensor placement;
b. choice of sensor (e.g., humidity and temperature sensor);
c. representativeness;
d. cloud detection only reported in cloud bases were below 12,000 ft;
e. 2-minute delay in response when the first clouds become present;

f. it cannot report tornadoes, ice pellets, drizzle, snowfall, snow depth, or blowing obstructions;
g. it cannot provide weather on a horizon to horizon basis; and
h. power outages.

Temperatures measured by an ASOS unit placed at separate locations at an airport differed by ≈2 °F, while a standard liquid-in-glass thermometer yielded consistent temperatures (Guttman and Baker, 1996); when temperature instruments were placed further apart on a cross bar at a given location, they indicated a colder (by a few tenths of a degree) than actual temperature. The original humidity sensor used was inexpensive and very prone to corrosion and degradation; this is to be replaced, if not already, with a LiCl or thin film capacitor. There is also a desire to make the wind sensors ice- and snow-free. Based on our discussion in Chapter 5, the sonic anemometer would make a strong candidate.

The aforementioned problems are easily remedied, and could have been avoided. The abilities to also measure frozen precipitation types, such as hail or sleet, as well as the ability to measure the intensity of the falling precipitation are in the planning as well. Perhaps the optical (laser) precipitation gauge might help out here. Nonetheless, once the optical precipitation gauge leaves the experimental stage it will provide a valuable tool to operational and research meteorologists.

In this chapter we discussed the characteristics of devices that made more than one of the fundamental parameters discussed in Chapters 3–7. Rawinsondes are continually improving, but one must be careful when analyzing large datasets derived from rawinsondes to ensure that their computations actually reveal the "true" environment and not that of the improvements. Automated weather systems were then discussed. The preliminary analysis of ASOS and conventional temperature data by Guttman and Baker (1996) shows that there is no straightforward, simple average correction that will adjust all of the data from conventional to ASOS. They also point out that any correction factors should include information on surface characteristics of the instrumentation sites, time of day, cloud amount, windspeed, persistence of conditions, and especially the influence of site characteristics on temperature measurement. This is similar for remote sensing data, except that the primary concern is to make sure the algorithm to "deconvolve" the data represents the "true" environment being sensed, and that the data used to "train' the algorithm is "truly" representative.

Even if automated weather observing stations put some people out of their jobs, there will always be a need to maintain these devices and to quality assure their performance, especially in the near future. Lastly, we were introduced to a system designed to meet the operational need of the National Weather Service. You are now ready to make a very good start in designing your measurement system to address your particular programmatic needs. The following exercises will help get you started.

## *Exercises*

1. Design a program to measure the windfield associated with a hurricane before and after it makes landfall.
2. Design a measurement program to measure the windfield associated with a tornado.
3. Design a measurement program to measure the indirect effect of aerosols on the radiative balance at the surface of the earth.
4. Design your own measurement program.

CHAPTER NINE

# Micrometeorological and Hydrometeorological Measurements

This chapter emphasizes the measurements of evaporation, evapotranspiration, soil moisture content, and soil heat and moisture fluxes near the surface since they are relevant to meteorology. The measurements of soil temperature and soil moisture flux are also included. The following has been derived from a number of references which include: Konstantinov (1966), Ward (1967), Sutton (1977), Dunne and Leopold (1978), Fritschen and Gay (1979), Brock (1984), Manning (1997) and Dr. U. Czapski, emeritus professor of atmospheric science at the State University of New York at Albany.

Several specialized disciplines of meteorology require special instrumentation or a more complex combination of instruments than have been described. Micrometeorology (including, and often synonymous with, boundary layer meteorology or turbulent transfer and diffusion) as well as hydrometeorology are two such disciplines.

Micrometeorology essentially deals with two closely interrelated problems:

1. the structure of the atmosphere close to the surface, depending on the nature of the surface (i.e., water or land, rough or smooth) and its temperature; and
2. the transport and diffusion of properties, such as heat, momentum, water vapor, and pollutants by the turbulent motions across, through, and within the surface.

These problems require the micrometeorologist to make two sets of field measurements:

1. profile measurements describing the mean fields of temperature, horizontal wind, water vapor concentration, and the concentration of atmospheric particulates or other quantities; and
2. precise measurements of the turbulent fluctuations of all three wind components, together with the simultaneous measurements of the fluctuations of temperature, water vapor concentrations, and the concentrations of other species.

The layer of air near the surface is commonly subdivided into three basic scales of boundaries, namely: skin or conductive, surface, and transition. The transition layer (between ≈50 to ≈100 m and ≈300 to ≈500 m above the surface) is where the vertical gradients in the mean horizontal windspeed, temperature, and the fluxes of heat, momentum, and other properties are variable. Viscosity is not important, but the surface drag force, temperature gradient, and coriolis force can be. The surface layer (between a few mm to a few cm and ≈50 to ≈100 m above the surface) is where the vertical gradients in the mean horizontal windspeed and the fluxes of heat, momentum, and other properties are constant. The nature of the surface and the vertical temperature gradient are important. The skin or conductive boundary layer is directly adjacent to the boundary where both vertical and horizontal components of motion are zero, and the fluxes of heat, momentum, and other properties are only due to molecular motions. Hence, the molecular conductivity and viscosity are in control. This boundary is quite shallow and its thickness (generally between a few millimeters to a few centimeters thick) depends on the temperature contrast between the air and the surface, as well as the roughness of the surface.

Given a solid surface, the momentum flux ends at the surface, but the heat flux at the surface can create a boundary layer with a potentially steep temperature gradient in the soil (or water) next to the boundary, since heat can be conducted from or into deep layers of the soil. Actually, if a steady state condition exists (i.e., same amount enters as leaves), then the heat flux (and sometimes that of water vapor) within the boundary layers of both media are equivalent. The flux within the soil is entirely conductive and can be determined by accurate measurements of the vertical temperature gradient close to the surface, and a knowledge of the heat conductivity of the soil. The aforementioned measurements may be obtained using a flux plate or thermistor device. The flux plate is similar to a net radiometer, and consists of a "sandwich" of two conductive sheets separated by a poorly conductive material. The temperature difference between the plates is determined by a thermopile or thermistors connected directly to or embedded within the surfaces. The thermistor device is simply a series of thermistors connected to a multiconductor cable. The main disadvantage of these measurements is the potential disruption of the structure of the soil material when deploying the instruments.

The hydrologist commonly applies the hydrologic or water balance equation,

$$\text{Inflow} - \text{Outflow} \pm \partial/\partial t\,(\text{Storage}) = 0 \qquad (9.1)$$

to quantify the water motion over, through, and across a land surface of a specific area for a specific time period. The area varies from a few square yards to the entire world and the time period from 10 minutes to on the order of 25 years, respectively. In contrast, the micrometeorologist studies the physical phenomena occurring on times scales of fractions of seconds to minutes to possibly hours. In Equation 9.1, precipitation is an example of Inflow. Evaporation losses, stream flow (Fig. 9.1), and seepage into the ground are examples of Outflow. Changes in groundwater level and soil moisture content are examples of changes in Storage.

**Fig. 9.1** *Stream flow is being measured at a number of points along the width of the stream using a current meter (similar to a propeller anemometer). The relationship of the water level to the volume flow of a stream through its width or cross-section establishes what is known as a rating curve. (Photo courtesy of Dr. Ulrich H. Czapski, Emeritus Professor of the State University of New York at Albany.)*

The geographical distribution or areal pattern of precipitation amount, its maximum and minimum values, the magnitude of the differences between maximum and minimum precipitation amounts, the temporal variations of the precipitation (including daily and seasonal scales), the predictability or recurrence of the temporal variations, and how much precipitation actually reaches the ground concern the hydrologist and meteorologist. The hydrologist further requires the precipitation measurement to be made at a height within 1 foot ($\approx$0.3 m) of the surface, and ideally at the surface. A "true" surface measurement of precipitation amount is difficult to obtain, primarily due to insplash (Ward, 1967).

The amount of precipitation that reaches the ground, known as interception, depends on the kind and density of the surface cover (i.e., vegetation, buildings, roads, pavement) upon which it falls. The intercepted precipitation sits on the surface, until it either falls to the ground, evaporates back into the atmosphere, or diffuses into the subground layers. The hydrologist would detail and quantify throughfall (i.e., water dripping through and from leaves to the ground), stemflow (i.e., water trickles along twigs and branches down the main trunk to the ground), runoff (i.e., water movement over the ground), infiltration (i.e., water moves into the ground; see Ward, 1967; Manning, 1997), and interception loss (i.e., evaporation of the water). For example, interception loss (I) is simply obtained from,

$$I = R - Rg - S. \tag{9.2}$$

Where R is the measured precipitation amount above the vegetation layer, Rg is the measured ground precipitation amount below the vegetation cover, and S is the stemflow, if applicable. "I" could vary from ≈5% for leafless trees, to 17% for bluegrass, to ≈45% for corn, to as high as 80% for spruce trees (e.g., Ward, 1967).

## *9.1 Evaporation and Evapotranspiration*

Theoretical or empirical relationships among air temperature, pressure, relative humidity, and windspeed may be used to estimate the evaporation of water from a surface. The estimates of evaporation from field measurements are primarily derived from the mass transfer, gravimetric, energy budget, mass budget, eddy flux, gradient flux, or combination mass transfer—energy budget method.

The mass transfer method is based on Dalton's law of partial pressures and involves measuring the vapor pressures and wind at the surface. The gravimetric measurement involves a direct measurement of water depth and windfield. The mass budget and energy budget methods are based on the conservation of mass and energy principles, and, according to Manning (1997), are not practically applied in the field. The flux methods require measurements that are made very frequently and at multiple near surface vertical atmospheric levels.

The gravimetric measurement method employs a small pan of a standard size developed in the U.S. for making operationally simple, direct estimates of evaporation near the surface. The National Weather Service has long used a galvanized sheet metal pan, termed the Class A pan (Fig. 9.2), that is 1.22 m in diameter and 0.25 m deep. The Class A pan is typically accompanied by an anemometer, a recording thermometer, a pyranometer, and a rain gauge. It is mounted on a wooden base as shown in Fig. 9.2 and is deployed with a preset water depth that is periodically monitored in conjunction with the accompanying meteorological data. The pan is protected with a fence to keep out animals, except birds. Manning (1997) points out that splashing birds can be a significant source of error, since covering the pan is obviously undesirable. The data from the pan is applied to larger water bodies, such as a lake or reservoir, by multiplying the measurement by a "multiplying" factor. The mean "multiplying" factor reported for Class A pans generally ranges between 0.5 and 0.82 with a mean range of 0.31 to 0.97 based on yearly data (Dunne and Leopold, 1978). Figure 9.3 provides an example of data from a pan study.

Konstantinov (1966) discusses the results of a very detailed, comprehensive USSR comparison study of the Class A and GGI-3000 evaporation pans, and a standard pan. The pans differ in area and depth, with the GGI-3000 pan 0.3 $m^2$ and 0.65 m deep. According to Konstantinov (1966), the report suggested that the differences between the two pans were mostly due to the way the pans were mounted. The class A pan is mounted on a wooden support (Fig. 9.2) that disturbs the flow, and it also absorbs an additional amount of heat that participates in the heat exchange, whereas the GGI-3000 was installed in the ground. The evaporation amounts (mm/day) recorded by evaporation pans of different sizes installed in the ground during the

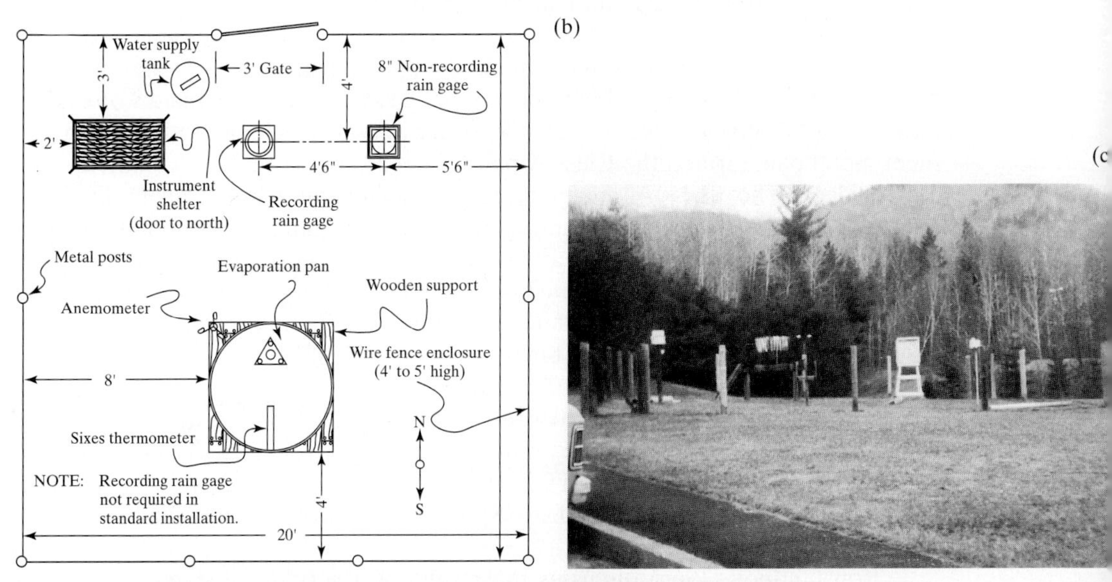


**Fig. 9.2** *(a) A National Weather Service galvanized sheet metal pan it termed the Class A pan (Photo courtesy of Qualimetrics, Inc.). (b) A layout plan for a standard evaporation pan station. (After U.S. Dept. of Commerce, 1990; Weather Station Handbook—an interagency guide for wildland managers, Nat. Wildfire Coord. Grp., PMS 426-2, NFES 2140). (c) An actual pan station site. Note the pan is located on the left side of the photo behind the non-recording gage. (Photo courtesy of Dr. Ulrich H. Czapski, Emeritus Professor of the State University of New York at Albany)*

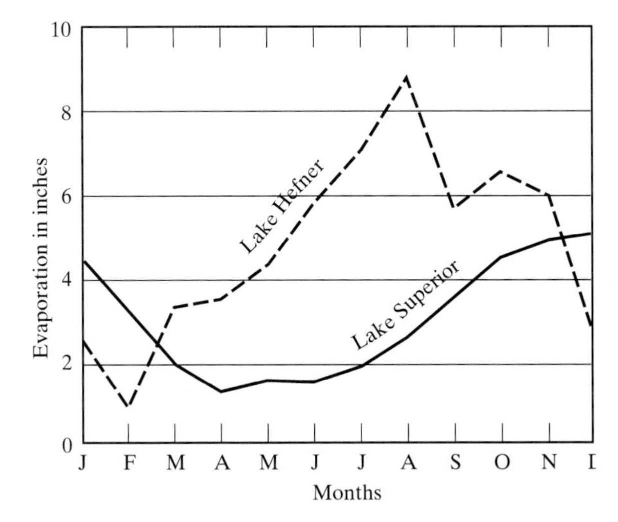


**Fig. 9.3** *An example of the annual evaporation obtained from a pan study. (Adapted from J. C. Manning, Applied Principles of Hydrology, 3rd ed., New York: Simon and Schuster, 1997)*

**Table 9.1** *Evaporation amounts (mm/day) recorded by evaporation pans of different sizes installed in the ground at Valdai Hydrological Research Laboratory during 1952–1957, (After Konstantinov, 1966)*

| | | *Evaporation Pans*[a] | |
|---|---|---|---|
| *Month* | *Standard Pan*[a] | *Class A* | *GGI-3000* |
| May | 2.3 | 2.1 | 2.3 |
| June | 3.0 | 3.0 | 3.1 |
| July | 2.9 | 2.7 | 2.9 |
| August | 2.3 | 2.0 | 2.3 |
| September | 1.4 | 1.2 | 1.4 |

[a]The dimensions of these pans are the same as those in the text.

summer months of part of the USSR study are shown in Table 9.1. The results of Table 9.1 show a better agreement compared to when the pans were just placed on the surface, since they are both measuring at the same level and any differences are due to their sizes, for example. Smaller pans have a higher measurement of the evaporation rate than do larger pans assuming a constant velocity and humidity as shown in Fig. 9.4 under different relative humidities. Thus the Class A pan should yield lower values than the GGI-3000 pan given that they are co-located and exposed to the same mean windspeed and relative humidity.

The mass transfer method for calculating evaporation is based on the assumption that the windspeed and the vapor pressure difference between the water surface and the atmosphere control the amount of water evaporated, $E_{evap}$, and is approximated as (Dunne and Leopold, 1978)

$$E_{evap} = N f(u)(e_{sw} - e), \qquad (9.3a)$$

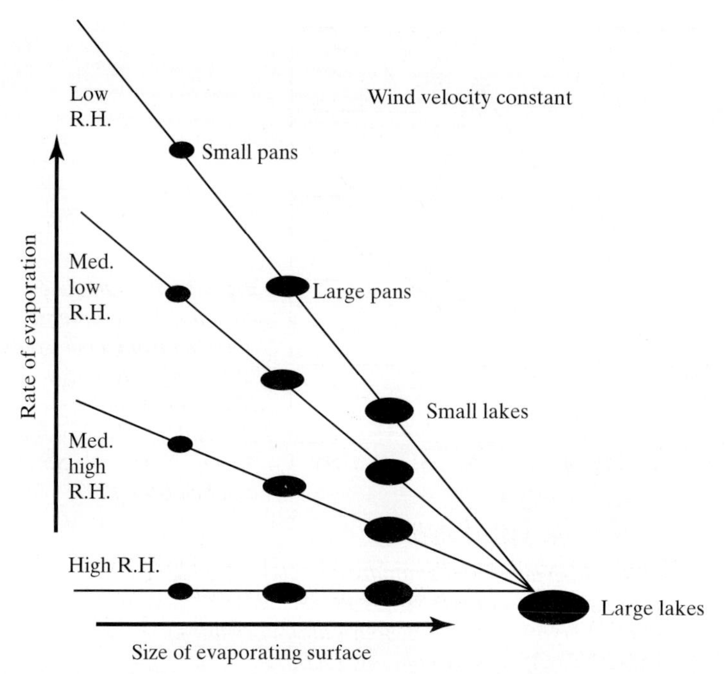


**Fig. 9.4** *The relationship between the rate of evaporation, relative humidity, and the size (area) of the evaporating surface. (Adapted from Ward, 1967.)*

where $e_{sw}$ is the water vapor pressure of the water surface, e is the water vapor pressure of the air, f(u) is a function of windspeed, u, and N is the mass transfer coefficient obtained from

$$N = 0.000169\, A_{wb}^{-0.05} \tag{9.3b}$$

in the arid southwestern U.S, where $A_{wb}$ is the area of the water body with units of $km^2$ and N has units of [mb (cm $day^{-1}$)/(km $day^{-1}$). N depends on the height of the measurements of u and e, turbulence characteristics of the airflow (also referred to as the stability regime), surface roughness, surface vegetation, surface permeability, and soil moisture (e.g., Brock, 1984). The reader is cautioned to keep track of the units when comparing mass transfer coefficients. Equation 9.3a assumes that there is no significant net change in the water surface elevation (i.e., no net inflow or net outflow) or no net groundwater movement below the surface, and it yields good results when applied in a semi-empirical fashion (Dunne and Leopold, 1978). For example see Fig. 9.5 which shows the measured evaporation rate plotted against the product of the mean windspeed times the change in vapor pressure, ($e_s$—e), with both windspeed and vapor pressure obtained at the same level near the ground (i.e., 2 m).

The energy budget method simply involves the energy balance

$$Q_s - Q_{rs} - Q_{lw} - Q_h - Q_e + Q_v - Q_{ve} = Q_\Theta \tag{9.4a}$$

where $Q_s$ is the energy per unit area from incoming solar radiation; $Q_{rs}$ is the energy per unit area from reflected solar radiation and may be expressed in terms of the

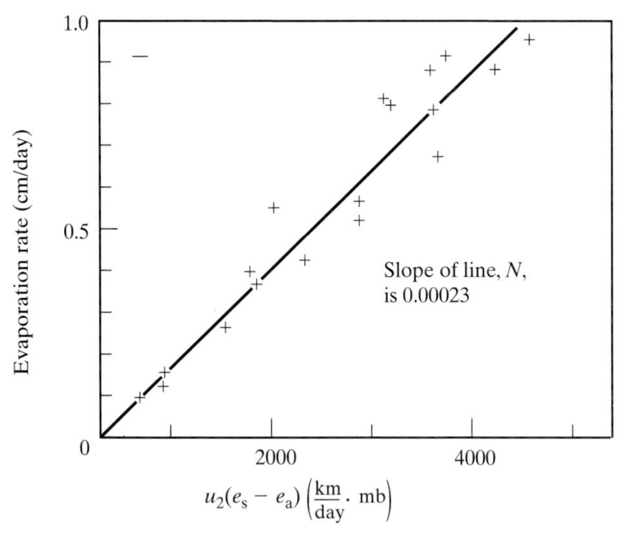


**Fig. 9.5** *The evaporation rate plotted against the product of the mean windspeed at a level near the ground (i.e., 2 m above ground level, AGL) times the change in vapor pressure, e, determined at the same near ground level as the mean windspeed. (After Dunne and Leopold, 1978)*

surface albedo, α, as $Q_{rs} = \alpha Q_s$; $Q_{lw}$ is the energy per unit area from the net long-wave radiation originating from the water body into the atmosphere; $Q_h$ is the energy per unit area from the sensible heat transferred by turbulent exchange from the water body to the atmosphere (i.e., sensible heat flux); $Q_e$ is the energy per unit area required to evaporate the water (i.e., the latent heat of vaporization flux); $Q_v$ is the energy per unit area from all water flows advecting into the particular water body; $Q_{ve}$ is the energy per unit area advecting out of the particular water body due to evaporation, and

$$Q_{ve} = Q_e c (T_{sw} - T_{st}) L^{-1} \tag{9.4b}$$

where $T_{sw}$ is the temperature of the water surface, $T_{st}$ is an arbitrarily chosen base temperature of 0 °C, L is the latent heat of vaporization, and c is the specific heat of water; and $Q_\theta$ is the change in the stored energy per unit area in the water body (i.e., below the surface). The expression for the amount of water evaporated may be obtained from mathematically manipulating Equations 9.4,

$$E_{evap} = (Q_s - Q_{rs} - Q_{lw} + Q_v - Q_\theta)\{\rho[c(T_{sw} - T_{st})] + [L(1 + R_{Bn})\}^{-1} \tag{9.5}$$

where $R_{Bn}$ is the Bowen ratio and $R_{Bn} = Q_h/Q_e = A(P/1000)\left(\frac{T_{sw} - T_{st}}{e_{sw} - e_{st}}\right)^{-1}$, A is the psychrometric constant (see Appendix D), $T_{sw}$ and $T_{st}$ have units of °C in this equation, P is air pressure with the same units as e (mb in this case), and all other symbols are as previously defined. Recall net radiation; here it equals $Q_s - Q_{rs} - Q_{lw}$ and is measurable with a single instrument (i.e., a net radiometer), calculated, or each term measured separately. The temperature and water vapor pressures are measured according to the methods discussed in previous chapters. $Q_v$ and $Q_\theta$ are obtained by making measurements of the temperature and volume of inflow, outflow, and water stored in the water body. Equation 9.5 neglects the heat from the precipitation hitting the ground, chemical–biological energy exchanges, and the kinetic heat dissipated in breaking waves (assuming a water body), since they are negligible compared to the terms shown (Brock, 1984).

The mass budget method is essentially the same as the energy budget equation, except

$$E_{evap} = \text{Inflow} + \text{Precipitation} - \text{Outflow} \pm \partial/\partial t\,(\text{Storage}). \tag{9.6}$$

The eddy flux method is simply based on the principle that turbulent air motions (or turbulent eddies) are more effective than diffusion in transporting moist air from a surface into the atmosphere. The evaporation of water from a surface under ideal conditions (i.e., the surface is large and homogeneous, and steady state conditions exist) is equal to a constant vertical transport of water vapor. Consequently, this method applies within the surface layer (see above). Thus, $E_{evap}$ may be estimated by measuring the covariance between the vertical component of the windfield, w, and the concentration of the water vapor, q (see Chapter 6),

$$E_{evap} = \rho <wq> = \rho K_e \partial q/\partial Z \tag{9.7}$$

where $< >$ indicates a time average, and all other quantities are as previously defined. The flux of water vapor from a homogeneous surface under steady state conditions may also be expressed in terms of the vertical gradient of the water vapor concentration, and $K_e$ is the eddy exchange coefficient or the eddy diffusivity, and Z is the height. The $\partial q/\partial Z$ portion of Equation 9.7 is the basis for the gradient flux method. Given approximate neutral atmospheric stability conditions and/or close to the surface, it is possible to make the assumption that $K_e$ for water vapor is similar to that for momentum, thereby implying that $K_e \approx K_m$, where $K_m$ is the eddy viscosity. Given the latter, then the $E_{evap}$ can be measured from the ratio of the gradients of the water vapor to that of the windspeed, u. Thornthwaite and Holzman (1939) were the first to derive an equation for evaporation based on the gradients of water vapor and windspeed, that was later generalized by Holzman (1943) and tested by others (e.g., Pasquil, 1950, and Rider, 1957). Equation 9.7 takes the following form assuming adiabatic gradient conditions, a logarithmic wind profile (e.g., Sutton, 1977), and, for simplicity, the vapor concentration, temperature and windfield are measured at 2 m and 8 m above ground level (AGL)

$$E_{evap} = \rho k^2 (u_8 - u_2)(q_2 - q_8)[(\ln(8/2))^2]^{-1} \tag{9.8}$$

where k is von Karman's constant, the subscripts indicate the levels of the measurements, and other symbols are as previously defined.

The measurements required for this method have been discussed in previous chapters. However, the instruments may have longer response times in this case as long as good accuracy is maintained. The mass transfer and energy budget methods are usually combined (Penman, 1961, 1967) to yield an empirical evaporation formula that can be modified to yield evapotranspiration.

Evapotranspiration refers to the return of moisture to the atmosphere from vegetated areas, total evaporation, and consumptive use (in agriculture, see Ward, 1967; Manning, 1997) which includes the water to make plant tissue. Some hydrology texts use consumptive use and evapotranspiration interchangeably. Evapotranspiration returns about two third of the annual precipitation to the atmosphere on the global scale (Manning, 1997), leaving the rest to flow into streams or into the ground. Evapotranspiration will be higher in arid regions compared to tropical regions. The reader is warned to pay attention to the evapotranspiration definition, since evapotranspiration is also split into two semantically different terms, namely actual evapotranspiration and potential evapotranspiration. The potential evapotranspiration is simply defined as the amount of moisture that will continuously meet the demands of transpiring vegetation (Ward, 1967), or as the total water loss from an extensive vegetated area with 100% coverage that is never short of water (Manning, 1997). Please see Ward (1967), Dunne and Leopold (1978), Manning (1997), and others for additional details.

Measurements of evapotranspiration are made in a similar manner as most of the techniques described for measuring evaporation. Evapotranspiration is indirectly measured using the mass transfer method (Equation 9.3), the mass budget method (Equation 9.6), or using a gravimetric device known as a lysimeter (Fig. 9.6). A neutron soil probe (Fig. 9.7) may be used to measure evapotranspiration as well as the moisture content of the soil at various depths below the surface as a neutron source is lowered down a tube. For example, using the mass budget method,

$$\text{ETr} = \text{Input from stream and canal flow} + \text{P} - \text{E} - \text{R} \pm \partial/\partial \text{t}\,(\text{Storage}), \quad (9.9)$$

where P is precipitation, E is evaporation, R is runoff, and the storage refers to surface lakes, reservoirs, soil water, and groundwater. Precipitation and runoff are both measured over a year or longer so that the storage term has a negligible value and may be ignored. The input from streams, canals and underground reservoirs or aquifers (Czapski, personal communication spring, 1997) are important in arid or semi-arid regions.

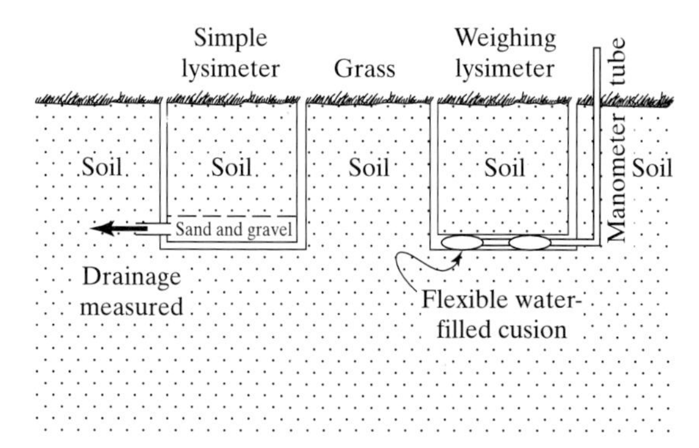


**Fig. 9.6** *General types of lysimeters. Note that in the simple lysimeter of drainage type the inflow into the soil is measured. (Adapted from J. C. Manning, Applied Principles of Hydrology, 3rd ed., New York: Simon and Schuster, 1997)*

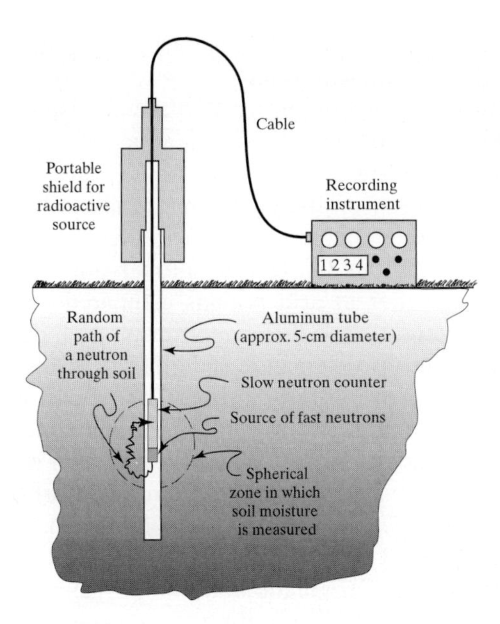


**Fig. 9.7** *A neutron soil probe. A radioactive source is lowered down the tube to a desired depth where it emits fast neutrons that follow random paths through the soil and are slowed by collision with hydrogen atoms of soil water. The neutrons slowed by the hydrogen atoms of water then migrate through the counter. The number counted per unit time represents a measure of the amount of water in the soil within the spherical volume indicated. (After Dunne and Leopold, 1978)*

## 9.2 *Soil Moisture and Temperature*

The moisture content of the soil varies with depth after water enters, or infiltrates, the soil as exemplified in Fig. 9.8. The neutron probe method as indicated above is capable of measuring the soil moisture content with depth. Figure 9.9 shows an example of the installation of tubes in which neutron sources are inserted. The insertion of the tubes may disturb the soil characteristics close to the tubes. The neutron probe measurements are relatively quick and do not themselves disturb the subsurface environment (Ward, 1967). They sample a volume of soil that is at least 1 $ft^3$ which leads to a relatively crude soil moisture profile. The soil moisture content may also be measured using a number of moisture transducers, such as a soil moisture reflectometer (Fig. 9.8), buried at desired intervals within the soil, and indirectly by the water balance method (e.g., Thornthwaite and Mather, 1954, 1955).

The temperature of the soil also varies with depth and an example of that variation is shown in Fig. 9.10. The measurement of temperature with depth may be done directly by placing thermistors that protrude from the sides of hollow tubes at predetermined intervals, with their wiring inside the tube leading up to the instrument above the surface. The soil temperature may be indirectly measured using a variant of the earth's radiation measuring devices, (e.g., see Fig. 9.10), or soil heat transducers placed at various depths below the surface. The direct approach should prove to be the most prudent as long as the sensor is placed in homogeneous soil, and conduction errors are minimal and due to the soil condition. Naturally this is the case for any of the soil transducer measurements.

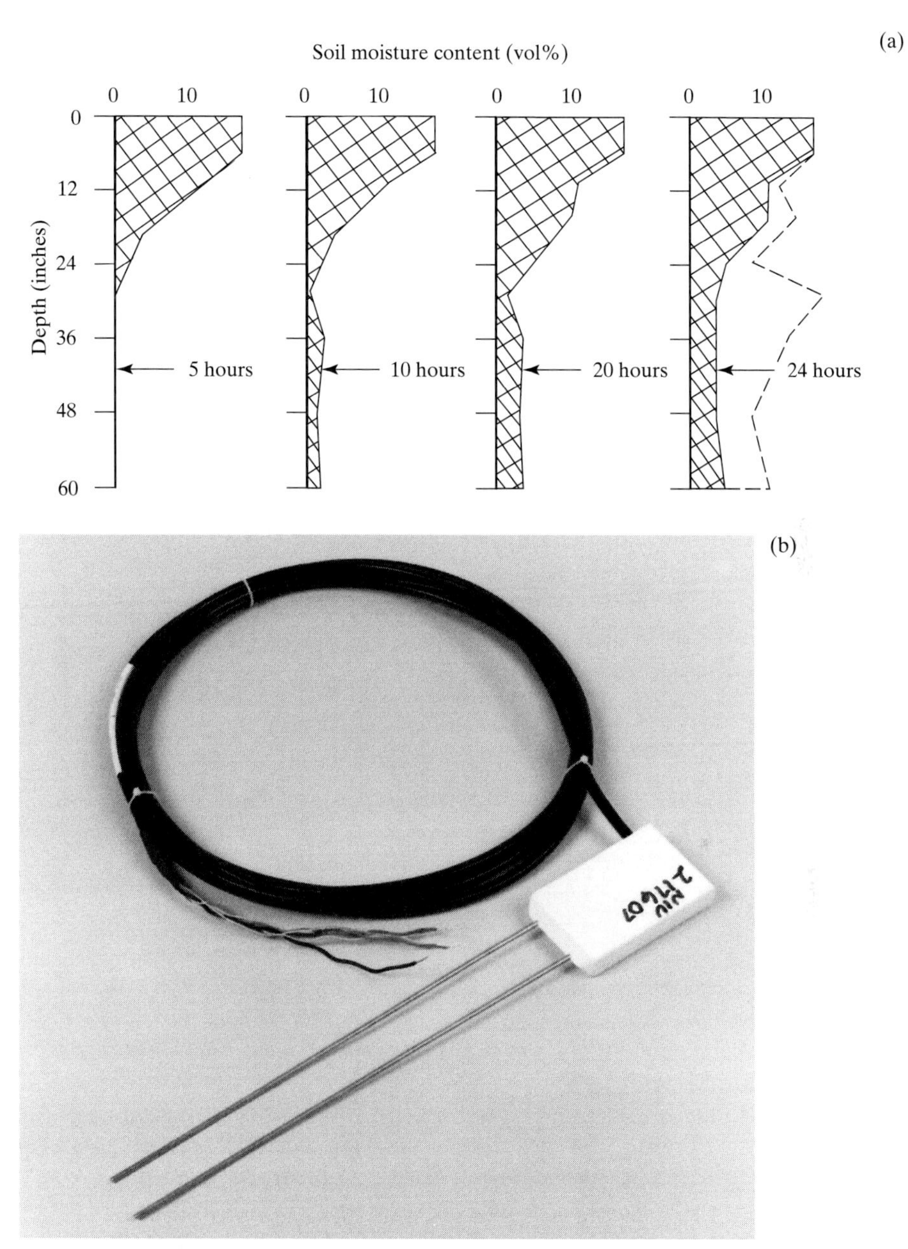


**Fig. 9.8** *(a) An example of the soil moisture content versus depth. (After Ward, 1967.) (b) A soil water reflectometer from Campbell Scientific, Inc. (Photo courtesy of Jie Song, Northern Illinois University)*

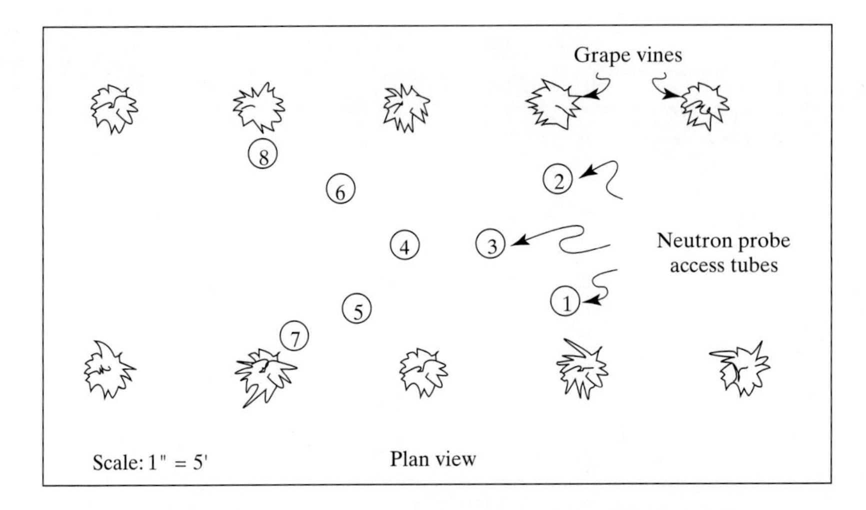


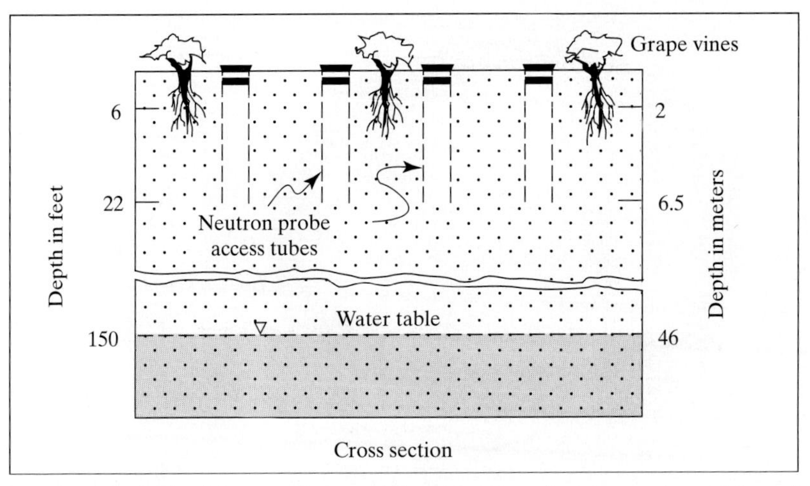


**Fig. 9.9** *Example of the installation of the neutron probe access tubes. (Adapted from J. C. Manning, Applied Principles of Hydrology, 3rd ed., New York: Simon and Schuster, 1997)*

## 9.3 Soil Heat and Moisture Fluxes

The concepts developed for transfer of heat and water vapor in the atmosphere apply to the transfer of these quantities in the soil. The transfer of heat and moisture within the atmosphere and even a water body is more efficient than that in the soil (Fleagle and Businger, 1980).

The heat flux from the soil, $Q'$, is typically very small except during spring and fall. It may be calculated near the surface provided you have (i) a near surface temperature

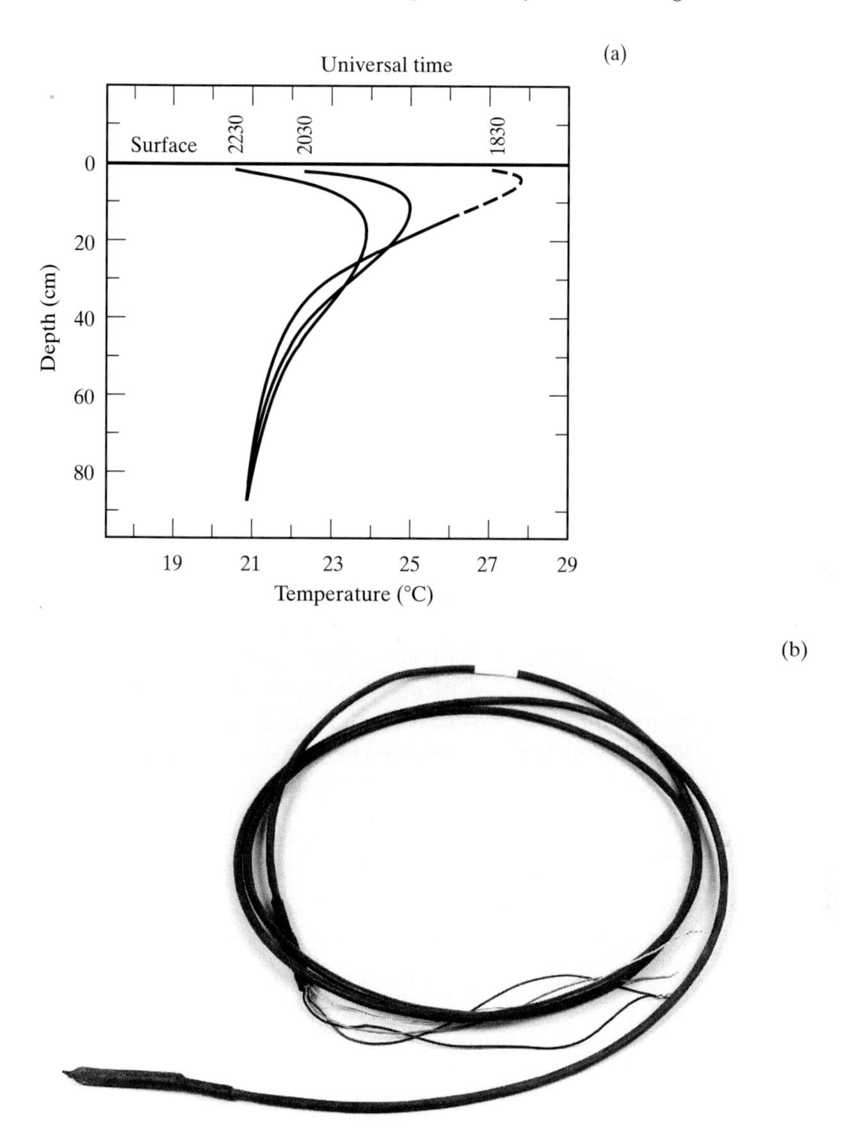


**Fig. 9.10** *(a) An example of the variation of temperature versus depth beneath the surface in O'Neill, Nebraska on 8 August 1953. (After Fleagle and Businger, 1979.) (b) Soil temperature probe #107 from Campbell Scientific, Inc. (Photo courtesy of Jie Song, Northern Illinois University.)*

gradient and know the thermal conductivity of the soil, or (ii) the temporal changes in the thermal conductivity and heat capacity of the soil.

$$Q' = [\sigma T_g^4] - [k_g(\partial T_{g,ave}/\partial Z')] - [\rho c_p K_H(\partial T_{ave}/\partial Z + \Gamma)] \tag{9.10}$$

where $\sigma$ is the Stefan-Boltzman constant, $T_g$ is the temperature of the ground surface (°K), $k_g$ is the thermal conductivity of the earth at the surface, $\partial T_{g,ave}/\partial Z'$ is the temperature gradient below the surface of the earth but as close as possible to the surface, $c_p$ is the specific heat capacity at constant pressure, $K_H$ is the eddy thermal diffusion coefficient of the air near the surface, $\partial T_{ave}/\partial Z$ is the mean temperature gradient in the layer $\partial Z$, and $\Gamma$ is the adiabatic lapse rate over the same $\partial Z$. $k_g$ ranges from $\approx 2$ W m$^{-1}$ m$^{-1}$ °C$^{-1}$ for wet soil or ice to $\approx 0.4$ for dry sand and $\approx 0.1$ for fresh snow. $k_g$ is also equal to the product of the density of the ground material times the heat capacity of the ground ($c_s$). The third term in Equation 9.10 is Brunt's expression for heat flux and is essentially the starting point for K theory used in micrometeorology (e.g., Sutton, 1977). The thermal conductivity or thermal diffusivity and the heat capacity of the soil are functions of soil moisture concentration or content, and are often unavailable.

The soil heat flux is simply determined by using a soil heat transducer (Fig. 9.11a) which is a direct measurement when used for this application because the flux is the desired quantity and not the temperature. The heat and fluxes from the surface into the atmosphere may be obtained by the gradient flux method using wind, temperature, and humidity profilers or individual measuring devices mounted in the vertical above and close to the surface (Fig. 9.11b). The latter profilers or vertically displaced measuring devices have been discussed in previous chapters.

The soil heat transducer, $G_T$, translates a heat flow through it into a temperature gradient across the transducer (Fritschen and Gay, 1979),

$$G_T = k_H(T_{top} - T_{bot})\,l^{-1} \tag{9.11}$$

where $k_H$ is the thermal conductivity, $T_{top}$ and $T_{bot}$ are the temperatures at the top and bottom of the transducer, and l is the thickness of the transducer. The thermal conductivity varies from 0.8 to 2.2 W m$^{-1}$ K$^{-1}$ in sand and clay soils with water contents of 8 to 20% by volume.

The soil heat flux out of the surface, $G_o$, derived from a transducer placed below the surface and concurrent temperature measurements may be expressed as

$$G_o = -[K_H(\partial T/\Delta Z')] - [c_s \Delta Z'(\partial T'_{ave}/\partial t)] \tag{9.12}$$

where $T'_{ave}$ is the average temperature of the soil layer, and $c_s$ is the heat capacity of the soil in the layer $\Delta Z'$ and has a nominal value of $\approx 2.9$ MJ m$^{-3}$ K$^{-1}$, with a value of $\approx 1.7$ MJ m$^{-3}$ K$^{-1}$ for very dry soil. The first term in Equation 9.12 is essentially measured by the heat transducer, whereas the second term is the change in heat energy stored above the transducer.

The number of soil flux sampling locations required depends on the project and the experimental site. There should be at least five soil flux measurements made per experimental site. The number of flux measurements depends on the surface conditions with fewer measurements required if the surface conditions are

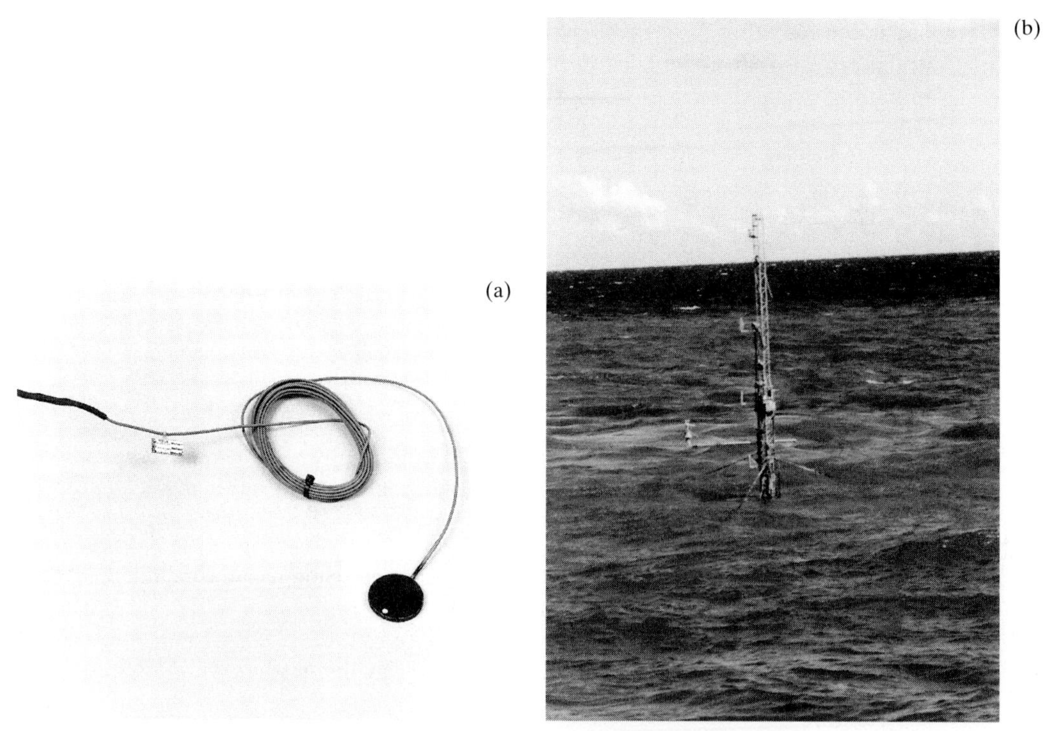

**Fig. 9.11** *(a) The soil heat flux is measureable with the soil heat flux probe from Campbell Scientific, Inc. The tag indicates that this probe has a calibration of 45.1 $Wm^{-1}/mV$. (Photo courtesy of Jie Song, Northern Illinois University) An instrumented tower with multiple level measurements is sometimes placed over a water or land surface to determine the heat transferred from it. (b) A micrometeorological research tower is situated in approximately 20 ft. of water off 9 mile point Nuclear Power Plant East of Oswego/ Lake Ontario for the measurement of heat transfer from the thermal plume. (Photo courtesy of Dr. Ulrich H. Czapski, Emeritus Professor of the State University New York at Albany)*

homogeneous (i.e., bare soil, dense grass, or uniform dense vegetation), but no less than three.

The soil heat transducer potentially yields errors up to 50%. Higher accuracies are possible, but the sensitivity usually suffers (Fritschen and Gay, 1979). Errors can be minimized by decreasing the thickness of the transducer and matching the thermal conductivity of the device with that of the soil being measured (Fritschen and Gay, 1979). The soil heat flux transducers should not be placed near the surface (Tanner, 1960; Fritschen and Gay, 1979), and Tanner (1960) suggested a depth of 5 to 10 cm below the soil surface.

In this chapter we briefly tapped the tip of the "iceberg" in micrometeorological and hydrometeorological measurements and relevant theory. Actually the measurements of primary interest to most meteorologists are variations of those discussed in previous chapters. The measurements of the actual values and the fluxes of heat and moisture are especially important for improving Global Climate Model simulations of hydrologic processes (e.g., Lau et al., 1996) as well as regional scale simulations (e.g., Meischner et al., 1993).

## Exercises

1. Class A and GGI-3000 pans are deployed in a typical manner with the Class A on a wooden support (made of 2" by 4" wooden studs) and the GGI-3000 on the ground.
   a. What is the approximate height of the evaporation measurement from each pan, and in which of the basic scale layers (e.g., skin, surface, transition) does each measurement fall?
   b. Based on a., determine which pan is expected to have a higher evaporation rate and why.
2. Expose a dish of water to a constant airstream from a fan. Place a psychrometer upwind, at the water surface and downwind of the dish. In addition place an anemometer near the water surface of the dish.
   What do the measurements indicate? Discuss your data.
   *Note*: If you do not have access to the measuring devices, choose your conditions and then answer the question.

# Bibliography

Ahr, M., A.I. Flossmann, and H.R. Pruppacher, 1989. On the effect of the chemical composition of atmospheric aerosol particles on nucleation scavenging and the formation of a cloud interstitial aerosol. *J. Atmos. Chemistry,* **9,** 465–478.

Atkins, P.W., 1986. *Physical Chemistry,* 3rd Ed. New York: W.H. Freeman and Co., 857 pp.

Atlas, D., ed., 1989. *Radar Meteorology.* Boston: American Meteorological Society.

Atlas, R., R.N. Hoffman, S.C. Bloom, J.C. Jusem, and J. Ardizzone, 1996. A multiyear global surface wind velocity dataset using SSM/I wind observations. *Bull. Amer. Meteorol. Soc.,* **77,** 869–888.

Bachalo, W.D., and M.J. Houser, 1984. Phase/Doppler spray analyzer for simultaneous measurements of drop size and velocity distributions. *Optical Engineering,* **23(5),** 583–590.

Baker, W.E., G.D. Emmitt, F. Robertson, R.M. Atlas, J.E. Molinari, D.A. Bowdle, J. Paegle, R.M. Hardesty, R. T. Menzies, T.N. Krishnamurti, R.A. Brown, M.J. Post, J.R. Anderson, A.C. Lorenc, and J. McElroy, 1995. Lidar-measured winds from space: A key component for weather and climate prediction. *Bull. Amer. Meteorol. Soc.,* **76,** 869–888.

Battan, L.J., 1973. *Radar Observation of the Atmosphere.* Chicago: Univ. of Chicago Press, 324 pp.

Baumgardner, D., 1983. An analysis and comparison of five water droplet measuring instruments. J. *Climate Appl. Meteorol.,* **22,** 891–910.

Baumgärtner, F., A. Zanier, and K. Krebs, 1989. On the mechanical stability of hollow NaCl particles. J. *Aerosol Sci.,* **20,** 883–886.

Beswick, K.M., K.J. Hargreaves, M.W. Gallagher, T.W. Choularton, and D. Fowler, 1991. Size-resolved measurements of cloud droplet deposition velocity to the forest canopy using an eddy correlation technique. *Q. J. R. Meteorol. Soc.*, **117**, 623–645.

Blanchard, D.C., and A.T. Spencer, 1964. Condensation nuclei and the crystallization of saline drops. J. *Atmos. Sci.,* **21,** 182–186.

Booth, C.R., T.B. Lucas, T. Mestechkina, J.R. Tusson IV, D.A. Neuschuler, and J.H. Morrow, 1995. NSF Polar Programs UV Spectroradiometer Network 1993–1994 Operations Report. (Available from Biospherical Instruments, Inc. 5340 Riley St., San Diego, CA. 92110–2621.) 166 pp.

Brazier, C.E., 1914. Recherches Expérimentales sur les Moulinets Anémométriques. *Annales du Bureau Central Météorologique de France*. pp. 157–300.

British Meteorological Office, Air Ministry, (BMOAM), 1961. *Handbook of Meteorological Instruments*. Her Majesty's Stationery Office, London, 500 pp.

Brock, F., ed., 1984. Instructor's Handbook on Meteorological Instrumentation. NCAR Technical Note, Boulder, 330 pp.

Businger, S., S.R. Chiswell, M. Bevis, J. Duan, R.A. Anthes, C. Rocken, R.H. Ware, M. Exner, T. VanHove and F. Solheim, 1996. The promise of GPS in atmospheric monitoring. *Bull. Amer. Meteorol. Soc.,* **77,** 5–18.

Chang, A.T., and T.T. Wilheit, 1979. Remote sensing of atmospheric water vapor, liquid water, and wind speed at the ocean surface by passive microwave techniques from the Nimbus 5 satellite. *Radio Sci.,* **14,** 793–802.

Chang, P.S., J.B. Mead, E.J. Knapp, G.A. Sadowy, R.E. Davis, and R.E. McIntosh, 1996. Polarimetric backscatter from fresh and metamorphic snowcover at millimeter wavelengths. *IEEE Trans. Antennas and Propagation,* **44,** 58–73.

Charlson, R.J., J.E. Lovelock, M.O. Andreae, and G.S. Warren, 1987. Oceanic phytoplankton, atmospheric sulphur, cloud albedo and climate. *Nature,* **326,** 655–661.

Cheng, R.J., 1970. Water drop freezing: Ejection of microdroplets. *Science,* **170,** 1395–1396.

Cheng, R.J., 1988. The generation of secondary marine aerosols: The crystallization of seawater droplets. In P. E. Wagner and G. Vali, eds., *Atmospheric Aerosols and Nucleation,* Springer-Verlag, pp. 589–592.

Cheng, R.J., 1989. Sulfate aerosols generation in the atmosphere: The evaporation of seawater droplets. *International Conference Global and Regional Environmental Atmospheric Chemistry,* May 3–10, Beijing, China, 192.

Cheng, R.J., D.C. Blanchard, and R.J. Cipriano, 1988. The formation of hollow sea salt particles from the evaporation of drops of sea water. *Atmos.* Res., **22,** 15–22.

Chiu, L., A. Chang, and J.E. Janowiak, 1993. Comparison of monthly rain rates derived from GPI and SSM/I using probability distribution functions. *J. Appl. Meteor.,* **32,** 323–334.

Choularton, T.W., D. Consterdine, I. Gardner, B. Gay, M. Hill, J. Latham, and I. Stromberg, 1986. Field studies of the optical and microphysical characteristics of clouds enveloping Great Dunn Fell. *Quart. J. Roy. Meterol. Soc.,* **112,** 131–148.

Chudamani, S., J.D. Spinhirne, and A.D. Clarke, 1996. Lidar aerosol backscatter cross sections in the 2 $\mu$m near-infrared wavelength region. *Appl. Optics,* **35,** 4812–4819.

Clements, R.H., 1964. Rainfall: Measurement, networks, data processing. Informal discussion of the Hydrological Group, *Proc. Inst. Civ. Engr.,* **29,** 258–260.

Cohen, A., M.A. Mahrer, and M. Litvak-Neumann, 1988. Dead sea evaporation rates in relation to aerosol number density variations. In P.V. Hobbs and M.P. McCormick, eds., *Aerosols and Climate.* A. Deepak Publ. Co., Hampton VA., pp. 21–30.

Cohn, S.A., and P.B. Chilson, 1995. NCAR workshop on multiple-receiver and multiple-frequency techniques for wind profiling. *Bull. Amer. Meteorol. Soc.,* **76,** 2474–2480.

Colby Jr, F.P., 1994. A Loran-based rawinsonde system for undergraduate meteorology programs. *Bull. Amer. Meteorol. Soc.,* **75,** 1027–1030.

Conner, M.D., and G.W. Petty, 1996. SSM/I brightness temperature deviations from gridded monthly means as a basis for over-land precipitation estimation. *8th Conf. on Satellite Meteorology and Oceanography.* 28 Jan.–2 Feb., 1996, Atlanta, GA, American Meteorological Society, Boston, MA., 001–004.

Cotton, W.R., and R.A. Anthes, 1989. *Storm and Cloud Dynamics.* New York: Academic Press, 881 pp.

Couch, R.H., C.W. Rowland, K.S. Ellis, M.P. Blythe, C.R. Regan, M.R. Koch, C.W. Antill, J.W. Cox, J.F. DeLorme, S.K. Crockett, R.W. Remus, J.C. Casas, and W.H. Hunt, 1991. Lidar in-space technology experiment: NASA's first in-space lidar system for atmospheric research. *J. Opt. Eng.,* **30,** 88–95.

Crescenti, G.H., 1997. A look back on two decades of Doppler Sodar comparison studies. *Bull. Amer. Meteorol. Soc.,* **77,** 651–673.

Cripps, D., and B. Abbott, 1997. The use of capacitance to detect icing. *J. Weather Modification,* **29,** 84–87.

Curry, J.A., C.D. Ardeel, L. Tian, 1990. Liquid water content and precipitation characteristics of stratiform clouds as inferred from satellite microwave measurements. *J. Geophys. Res.* **95D,** 16659–16671.

Czys, R.R., 1991. An appraisal of the microphysical nature and seedability of warm-based midwestern clouds at 10 °C. *J. Wea. Mod.,* **23,** 1–17.

Czys, R.R., and M.S. Petersen, 1992. A roughness detection technique for objectivity classifying drops and graupel in 2D image records. *J. Atmospheric and Oceanic Technol.,* **9,** 242–257.

Daube, B.C., R.C. Flagan, and M.R. Hoffmann, 1987. Active cloudwater collector. United States Patent No. 4,697,462. 6 October.

DeFelice, T. P., 1996. Variations in cloud condensation nuclei at Palmer Station Antarctica during February 1994. *Atmospheric Research,* **41,** 229–248.

DeFelice, T.P., and R.J. Cheng, 1998. On the phenomenon of nuclei enhancement during the evaporative stage of a cloud. *Atmos. Res.,* In press.

DeFelice, T. P., and V. K. Saxena, 1990. Mechanisms for the operation of three cloudwater collectors: Comparison of mountain-top results. *Atmospheric Research,* **25,** 277–292.

DeFelice, T.P., and V.K. Saxena, 1994. On the variation of cloud condensation nuclei in association with cloud systems at a mountain-top location. *Atmos. Res.,* **31,** 13–39.

Demoz, B.B., A.W. Huggins, J.A. Warburton, and R.L. Smith, 1993. Field performance of a spinning-reflector microwave radiometer. J. *Atmospheric and Oceanic Technol.,* **10,** 421-427.

Demoz, B.B., J.L. Collett Jr., and B.C. Daube Jr., 1996. On the Caltech active strand cloudwater collectors. *Atmospheric Research,* **41,** 47–62.

Dessens, H., 1946. Les noyaux e condensation de l'atmosphère. *C.R. Acad. Sci.,* **223,** 915–917.

Dinger, J.E., H.B. Howell, and T.A. Wojciechowski, 1970. On the source and composition of cloud nuclei in a subsident air mass over north Atlantic. *J. Atmos. Sci.,* **27,** 791–797.

Doesken, N.J., and A. Judson, 1996. *The snow booklet: A guide to the science, climatology and measurement of snow in the United States.* Colorado State University-ISBN 0-9651056–1-x. 84 pp. (Available through authors, Colorado State Climatology Office at Colorado State Univ. at Fort Collins).

Doviak, R.J., and D.S. Zrnic, 1993. *Doppler Radar and Weather Observations.* New York: Academic Press, 562 pp.

Dunne, T., and L.B. Leopold, 1978. *Water in Environmental Planning.* San Francisco: W.H. Freeman and Co., 818 pp.

Dye, J.E., and D. Baumgardner, 1984. Evaluation of the forward scattering spectrometer probe. Part I-Electronic and optical studies. *J. Atmospheric and Oceanic Technol.,* **1,** 329–344.

Ellingson, R.G., and W.J. Wiscombe, 1996. The spectral radiance experiment (SPECTRE): Project description and sample results. *Bull. Amer. Meteorol. Soc.,* **77,** 1967–1987.

Ellrod, G., and J.P. Nelson III, 1996. Assessment of GOES–8 imager data quality. 8th Conf. on Satellite Meteorology and Oceanography, 28 Jan.–2 Feb., Atlanta, GA, American Meteorological Society, Boston, MA.

Eskridge, R.E., O.A. Alduchov, I.V. Chernykh, Z. Panmao, A.C. Polansky, and S.R. Doty, 1995. A comprehensive aerological reference data set (CARDS): Rough and systematic errors. *Bull. Amer. Meteorol. Soc.,* **76,** 1759–1775.

Evans, K.F., and J. Vivekanandan, 1990. Multiparameter radar and microwave radiative transfer modelling of nonspherical atmospheric ice particles. *IEEE Trans. Geosci. Remote Sens.,* **28,** 423–427.

Ferraro, R.R., N.C. Grody, and G.F. Marks, 1994. Effects of surface conditions on rain identification using the SSM/I. *Remote Sens. Rev.,* **11,** 195–209.

Ferraro, R.R., F. Weng, N.C. Grody, and A. Basist, 1996. An eight-year (1987–1994) time series of rainfall, clouds, water vapor, snow cover and sea ice derived from SSM/I measurements. *Bull. Amer. Meteorol. Soc.,* **77,** 891–906.

Finkelstein, P.L., D.A. Mazzarella, T.J. Lockhart, W.J. King, and J.M. White, 1983. *Quality Assurance Handbook for Air Pollution Measurement Systems: Vol. IV, Meteorological Measurements.* **EPA–600/4–82–060.** Environmental Monitoring Systems Laboratory, Research Triangle Park, NC, 27711.

Fleagle, R.G., and J.A. Businger, 1980. *An Introduction to Atmospheric Physics.* New York: Academic Press, 432 pp.

Fritschen, L.J., and L.W. Gay, 1979. *Environmental Instrumentation.* New York: Springer-Verlag, 216 pp.

Fuchs, M., 1971. Data logging and scanning rate considerations in micrometeorological experiments: A discussion. *Agric. Meteorol.,* **9,** 285–286.

Fujiyoshi, Y., T. Endoh, T. Yamada, K. Tsuboki, Y. Tachibana, and G. Wakahama, 1990. Z-R relationship for snow. *J. Appl. Meteorol.,* **29,** 147–152.

Furger, M., C.D. Whiteman, and J.M. Wilczak, 1995. Uncertainty of boundary layer heat budgets computed from wind profiler-RASS networks. *Mon. Wea. Rev.,* **123,** 790–799.

Garand, L., G. Grassotti, J. Halle and G.L. Klein, 1992. On differences in radiosonde humidity: Reporting practices and their implications for numerical weather prediction and remote sensing. *Bull. Amer. Meteorol. Soc.,* **73,** 1417–1423.

Gedzelman, S.D., and J.R. Lawrence, 1982. The isotopic composition of cyclonic precipitation. *J. Appl. Meteorol.,* **21,** 1385–1404.

Gerber, H.E., 1980. Saturation hygrometer for the measurement of relative humidity between 95% and 105%. *J. Appl. Meteorol.,* **19,** 1196–1238.

Gertler, A.W., 1983. *A comparison study of fog collectors in the Los Angeles Basin.* Final Report. Coordinating Research Council, Inc., Atlanta, Georgia. No. CAPA-21-80(7-83), August 2. (Available from A.W. Gertler, Atmospheric Sciences Center, Desert Research Institute, P.O. Box 60220, Reno, Nevada 89506), 10 pp.

Ghan, S.J., K.E. Taylor, J.E. Penner, and D.J. Erickson III, 1990. Model test of CCN-cloud albedo climate forcing. *Geophys. Res. Lettrs.,* **17,** 607–610.

Giraytys, J., 1970. Can instruments be designed to be both accurate and representative? Summary of a panel discussion. *Meteor. Monogr.,* **2(33),** 437–443.

Goldsmith, J.E.M., S.E. Bisson, R.A. Ferrare, K.D. Evans, D.N. Whiteman, and S.H. Melfi, 1994. Raman lidar profiling of atmospheric water vapor: Simultaneous measurements with two collocated systems. *Bull. Amer. Meteorol. Soc.,* **75,** 975–994.

Goody, R.M., and Y.L. Yung, 1989. *Atmospheric Radiation: Theoretical Basis,* 2nd. ed., New York: Oxford Univ. Press.

Grantham, D.D., ed., 1995. Cloud modeling and data for defense simulation activities: "Emphasizing sufficient physical reality in simulating clouds." *Cloud Impacts on DOD operations and systems 1995 conference* (CIDOS-95). (Available from D.D. Grantham, Phillips Laboratory Directorate of Geophysics, Air Force Materiel Command, Hanscom Air Force Base, MA 01731-3010. PL-TR-95-2129, Environmental Research Papers, No. 1179.), 270 pp.

Grody, N.C., 1976. Remote sensing of atmospheric water from satellite using microwave radiometry. *IEEE Transactions Ant. Prop.,* AP-**24,** 155–162.

Grody, N.C., and A.N. Basist, 1996. Global identification of snowcover using SSM/I measurements. *IEEE Transactions on Geoscience and Remote Sensing,* **34,** 237–249.

Guirard, F.O., J. Howard, and D.C. Hogg, 1979. A dual-channel microwave radiometer for measurement of precipitable water vapor and liquid. *IEEE Transactions on Geoscience Electronics,* **GE-17(4),** 129–136.

Guttman, N.B., and C.B. Baker, 1996. Exploratory analysis of the difference between temperature observations recorded by ASOS and conventional methods. *Bul. Amer. Meteorol. Soc.,* **77,** 2865–2873.

Hallett, J., and S.C. Mossop, 1974. Production of secondary ice crystals during the riming process. *Nature,* **249,** 26–28.

Hegg, D.A., 1990. Heterogeneous production of cloud condensation nuclei in the marine atmosphere. *Geophys. Res. Lettrs.,* **17,** 2165–2168.

Hegg, D.A., 1991. Particle production in clouds. *Geophys. Res. Ltrs.,* **18,** 995–998.

Hegg, D.A., and P.V. Hobbs, 1979. The homogeneous oxidation of sulfur dioxide in cloud droplets. *Atmos. Environ.,* **13,** 981–987.

Hegg, D.A., and P.V. Hobbs, 1986. Sulfate and nitrate chemistry in cumuliform clouds. *Atmos. Environ.,* **20,** 901–909.

Hegg, D.A., P.V. Hobbs, and L.F. Radke, 1980. Observations of the modification of cloud condensation nuclei in wave clouds. *J. Rech. Atmos. (Atmos. Res.),* **14,** 217–222.

Hegg, D.A., L.F. Radke, and P.V. Hobbs, 1990. Particle production associated with marine clouds. *J. Geophys. Res.,* **95D,** 13917–13926.

Hering, S.V., D.L. Blumenthal, R.L. Brewer, A. Gertler, M. Hoffmann, J.A. Kadlecek, and K. Pettus, 1987. Field intercomparison of five types of fogwater collectors. *Environ. Sci. Technol.,* **21,** 654–663.

Heymsfield, A.J., and L.M. Miloshevich, 1989. Evaluation of liquid water measuring instruments in cold clouds sampled during fire. *J. Atmos. Oceanic Technol.,* **6,** 378–388.

Hill, G.E., 1990. *Radiosonde supercooled liquid water detector.* Final Report for U.S. Army CCREL, Hanover, NH. DACA 89–84-C–0005. 97 pp. [Available from ATEK Data Corp. 2300 Canyon Blvd., Boulder, CO 80302].

Hindman, E.E., 1987. A "cloud-gun" primer. *J. Atmos. Ocean. Technol.,* **4,** 736–741.

Hindman, E.E., P.A. Durkee, and W. Kuenning, 1985. Properties of aerosol particles detected by satellites in coastal regions. *J. Rech. Atmos. (Atmos. Res.),* **19,** 315–322.

Hinze, J.O., 1959. *Turbulence.* New York: McGraw-Hill, 586 pp.

Hoffer, T.E., and S.C. Mallen, 1970. Evaporation of contaminated and pure water droplets in a wind tunnel. *J. Atmos. Sci.,* **27,** 914–918.

Hogan, A.W., 1982. Estimation of sulfate deposition. *J. Appl. Meteorol.,* **21,** 1933–1936.

Holzman, B., 1943. *Ann. N.Y. Acad. Sci.,* **44,** 13.

Hood, R.E., R.W. Spencer, F.J. LaFontaine, and E.A. Smith, 1994. Simulation of future passive microwave satellite instruments using high resolution AMPR aircraft data. *Proc. AMS Satellite Conf.* AMS, 3.10, 1–4. (Available from R.E. Hood, NASA/MSFC, GHCC, ES43, Marshall Space Flight Center, AL 35812).

Hoppel, W.A., 1988. Nonprecipitating cloud cycles and gas-to-particle conversion. In P.V. Hobbs and M.P. McCormick, eds., *Aerosols and Climate.* A. Deepak Publ. Co., Hampton VA, pp. 9–19.

Hoppel, W.A., J.W. Fitzgerald, G.M. Frick, R.E. Larson, and E.J. Mack, 1990. Aerosol size distributions and optical properties found in the marine boundary layer over the Atlantic Ocean. *J. Geophys. Res.,* **95,** 3659–3686.

Houze Jr, R.R., 1993. *Cloud Dynamics.* New York: Academic Press, 573 pp.

Hucek, R., L. Stowe, and R. Joyce, 1996. Evaluating the design of an earth radiation budget instrument with system simulations. Part III: CERES-I diurnal sampling error. *J. Atmos. Oceanic Technol.,* **13,** 383–399.

Hudson, J. G., 1984. Cloud condensation nuclei measurements within clouds. *J. Clim. Appl. Meteorol.,* **23,** 42–51.

Huff, F.A. and J.C. Neill, 1957. *Rainfall relations on small areas in Illinois.* Bulletin 44. Illinois State Water Survey, Urbana. 61 pp.

Huffman, G.J., R.F. Adler, P. Arkin, A. Chang, R. Ferraro, A. Gruber, J. Janowiak, A. McNab, B. Rudolf, and U. Schneider, 1997. The global precipitation climatology project (GPCP) combined precipitation dataset. *Bul. Amer. Meteorol. Soc.,* **78,** 5–20.

Jacob, D.J., J.M. Waldman, M. Haghi, M.R. Hoffman, and R.C. Flagan, 1985. An instrument to collect fogwater for chemical analysis. *Rev. Sci. Instrum.,* **56,** 1291–1293.

Jones, A.S., and T.H. Vonder Haar, 1990. Passive microwave remote sensing of cloud liquid water over land regions. *J. Geophys. Res.,* **95D,** 16673–16683.

Kaimal, J.C., and J.E. Gaynor, 1983. The Boulder atmospheric observatory. *J. Climate Appl. Meteorol.,* **22,** 863–880.

Katz, U., and D. Miller, 1984. *Development and Evaluation of Cloud Water Collectors.* Final report to Electric Power Research Institute, Palo Alto, CA and to Coordinating Research Council, Inc., Atlanta, GA. Desert Research Institute, Reno, NV, April, 96 pp.

Kidder, S.Q., and T.H. Vonder Haar, 1977. Seasonal oceanic precipitation frequencies from Nimbus 5 microwave data. *J. Geophys. Res.,* **82,** 2083–2086.

Kidder, S.Q., and T.H. Vonder Haar, 1995. *Satellite Meteorology: An Introduction.* New York: Academic Press, 466 pp.

Kikuchi, K., and A.W. Hogan, 1976. Snow crystal observations in summer season at Admundsen-Scott South Pole Station, Antarctica. *Hokkaido Univ., Faculty of Science, Journal, Series 7,* **Geophys., 5(1),** 1–20.

King, L.V., 1914. On the convection of heat from small cylinders in a stream of fluid. *Philos. Trans. R. Soc. London Ser.,* **A214,** 373–432.

King, W.D., D.A. Parkin, and R.J. Handsworth, 1978. A hot-wire liquid water device having fully calculable response characteristics. *J. Appl. Meteorol.,* **17,** 1809–1813.

King, W.J., 1943. Measurement of high temperature in high velocity gas streams. *Am. Soc. Mech. Eng. Pap.,* **65,** 421–431.

Kondratyev, K. Ya., 1969. *Radiation in the Atmosphere.* New York: Academic Press.

Konstantinov, A.R., 1966. *Evaporation in Nature.* Israel Program for Scientific Translations, Ltd. (Available from U.S. Dept. Commerce, Clearinghouse for Federal Scientific and Technical Information, Springfield, VA 22151). 523 pp.

Kummerow, C. and L. Giglio, 1994a. A passive microwave technique for estimating rainfall and vertical structure information from space. Part I. algorithm description. *J. Appl. Meteorol.,* **33,** 3–18.

Kummerow, C. and L. Giglio, 1994b. A passive microwave technique for estimating rainfall and vertical structure information from space. Part II. applications to SSM/I data. *J. Appl. Meteorol.* **33,** 19–34.

Lally, V.E., 1985. Upper air in situ observing systems. In David D. Houghton, ed. *Handbook of Applied Meteorology.* New York: John Wiley & Sons, Inc., 352–360.

Lau, K.-M., J.H. Kim, and Y. Sud, 1996. Intercomparison of hydrologic processes in AMIP GCMs. *Bullet. Amer. Meteorol. Soc.,* **77,** 2209–2227.

Leong, K.H., 1981. Morphology of aerosol particles generated from evaporation of solution droplets. *J. Aerosol Sci.,* **12,** 417–435.

Letestu, S., 1966. *International Meteorological Tables.* World Meteorological Organization, WMO-no. 118. TP.94, Geneva, Switzerland, 1966; with amendments through July 1973.

Lin, N.-H., and V.K. Saxena, 1991. In-cloud scavenging and deposition of sulfates and nitrates: Case studies and parameterization. *Atmos. Environ.,* **25A,** 2301–2320.

Liou, K.-N., 1980. *An Introduction to Atmospheric Radiation.* New York: Academic Press.

List, R.J., 1961. Hailstones. *Bul. Amer. Meteorol. Soc.,* **42,** 452.

List, R.J., 1971. *Smithsonian Meteorological Tables.*

Lockyer, R., 1996. A Sonic Anemometer for General Meteorology. SENSORS, **13,** 3 pp. Peterborough, NH: Helmers Publishing Inc.

Lodge, J.P., 1994. *Methods of air sampling and analysis,* 3rd ed. Ann Arbor: CRC Press, 700 pp.

Lodge, J.P., and F. Baer, 1954. An experimental investigation of the shatter of salt products on crystallization. *J. Meteorol.,* **11,** 420–421.

Lojek, D., 1996. ASOS maintenance support system. *Preprints, 12th International Conference on Interactive Information Processing Systems,* Atlanta, GA, American Meteorological Society, Boston, MA.

Lovett, G.M., 1984. Rates and mechanisms of cloud water deposition to a subalpine balsam fir forest. *Atmos. Environ.,* **19,** 797–802.

Ludlum, F.H., 1980. *Clouds and Storms.* University Park: Pennsylvania State Univ. Press, 464 pp.

MacCready, Jr., P.B., 1965. Dynamic response characteristics of meteorological sensors. *Bull. Amer. Meteorol. Soc.,* **46,** 533–538.

McClain, E.P., W.G. Pichel, and C.C. Walton, 1985. Comparative performance of AVHRR-based multichannel sea surface temperatures. *J. Geophys. Res.,* **90,** 11587–11601.

McCormick, M.P., D.M. Winker, E.V. Browell, J.A. Coakley, C.S. Gardner, R.M. Hoff, G.S. Kent, S.H. Melfi, R.T. Menzies, C.M.R. Platt, D.A. Randall, and J.A. Reagan, 1993. Scientific investigations planned for the Lidar In-Space Technology Experiment (LITE). *Bul. Amer. Meteorol. Soc.,* **74,** 205–214.

McIllroy, I.C., 1955. C.S.I.R.O. Met. Phys. Technical Paper No. 3.

McIllroy, I.C., 1961. C.S.I.R.O. Met. Phys. Technical Paper No. 11.

McLaren, S., J. Kadlecek, A. Kadlecek, J. Spencer, and C. Conway, 1985. *Field intercomparison of cloud water collectors at Whiteface Mountain.* Report from the Whiteface Mountain field station, Wilmington, NY 12997. Atmospheric Science Research Center, 66 pp.

Manning, J.C., 1997. *Applied Principles of Hydrology.* Upper Saddle River, NJ: Prentice Hall, 276 pp.

Martin, D.W., and W.D. Scherer, 1973. Review of satellite rainfall estimation methods. *Bull. Amer. Meteorol. Soc.,* **54,** 661–674.

Matsumoto, T., P.B. Russell, C. Mina, W. VanArk, and V. Banta, 1987. Airborne tracking sunphotometer. *J. Atmosph. Ocean. Technol.,* **4,** 336–339.

Mayewski, P.A., M.J. Spencer, W. B. Lyons, and M.S. Twickler, 1987. Seasonal and spatial trends in south Greenland snow chemistry. *Atmos. Environ.,* **21,** 863–869.

Mazzeralla, D.A., 1972. An inventory of specifications for wind measuring instruments. *Bull. Amer. Meteorol. Soc.,* **53,** 860–871.

May, P., R. Strauch, and K. Moran, 1988. The altitude coverage of temperature measurements using RASS with wind profiler radars. *Geophys. Res. Lett.,* **15,** 1381–1384.

Meischner, P.F., M. Hagen, T. Hauf, D. Heimann, H. Höller, U. Schumann, W. Jaeschke, W. Mauser, and H.R. Pruppacher, 1993. The field project CLEOPATRA, May–July 1992 in Southern Germany. *Bullet. Amer. Meteorol. Soc.,* **74,** 401–412.

Middleton, W.E.K., and A.F. Spilhaus, 1953. *Meteorological Instrumentation.* Toronto: Univ. of Toronto Press, 286 pp.

Mitchell, D.L., 1995. Use of mass- and area- dimensional power laws for determining precipitation particle fallspeeds. *J. Atmos. Sci.,* **52,** 001–014. (Available from D.L. Mitchell, DRI, Reno, NV 89506.)

Mitra, S.K., J. Brinkmann, and H.R. Pruppacher, 1992. A wind tunnel study on the drop-to-particle conversion. *J. Aerosol Science,* **23,** 245–256.

Mohr, K.I., and E.J. Zipser, 1996. Defining mesoscale convective systems by their 85 GHz ice-scattering signatures. *Bull. Amer. Meteorol. Soc.,* **77,** 1179–1190.

NAPAP, National Acid Precipitation Assessment Program, 1987. *Interim Assessment: The Causes and Effects of Acid Deposition,* Washington, D.C.: J.L. Kulp and C.N. Herrick, eds., Vol. 1, U.S. Govt. Printing Office, 1–53.

Nappo, C.J., J.Y. Caneill, R.W. Furman, F.A. Gifford, J.C. Kaimal, M.L. Kramer, T.J. Lockhart, M.M. Pendergast, R.A. Pielke, D. Randerson, J.H. Shreffler, and J.C. Wyngaard, 1982. The workshop on the representativeness of meteorological observations, June 1981, Boulder, CO. *Bull. Amer. Meteorol. Soc.,* **63,** 761–764.

Neff, W.D., 1990. Remote sensing of atmospheric processes over complex terrain. *Atmospheric Processes over Complex Terrain, Meteorol. Monogr.,* No. **45,** Amer. Meteorol. Soc., 173–228.

OMEGA, 1992. *OMEGA Temperature Measurement Handbook and Encyclopedia,* Vol. 28, North American Edition, OMEGA Engineering, Stamford, CT, 1130 pp.

OMEGA, 1994. *The Infrared Temperature Handbook,* Vol. 29, North American Edition, OMEGA Engineering, Stamford, CT, 120 pp.

OMEGA, 1995. OMEGA Temperature Measurement Handbook and Encyclopedia, Vol. 29, North American Edition, OMEGA Engineering, Stamford, CT, 1140 pp.

Orville, H.D., 1996. A review of cloud modelling in weather modification. *Bull. Amer. Meteorol. Soc.,* **77,** 1535–1555.

Paine, L.C. and H.R. Farrah, 1965. Design and applications of high-performance dew-point hygrometer. *Humidity and Moisture* (eds. A. Wexler and R. E. Ruskin). New York: Reinhold Publishing Co. London: Chapman and Hall, Ltd., 174–188.

Pandis, S.N., and J.H. Seinfeld, 1989. Mathematical modeling of acid deposition due to radiation fog. *J. Geophys. Res.,* **94D,** 12911–12923.

Pani, E.A., and G.M. Jurica, 1989. A Z-R relationship for summertime convective clouds over west Texas. *J. Appl. Meteor.,* **28,** 1031–1037.

Pasquil, F., 1950. Some further consideration of the measurement and indirect evaluation of natural evaporation. *Q. J. Roy. Meteorol. Soc.,* **76,** 287–301.

Penman, H.L., 1961. Weather, plant and soil factors in hydrology. *Weather,* **16,** 207–219.

Penman, H.L., 1967. Evaporation from forests, a comparison of theory and observations. In W.E. Sopper and H.W. Lull, eds. *Forest Hydrology,* Oxford: Pergamon Press, 373–380.

Petty, G.W., 1994a. Physical retrievals of over-ocean rain rate from multichannel microwave imagery. Part I: Theoretical characteristics of normalized polarization and scattering indices. *Meteorol. Atmos. Phys.,* **54,** 79–100.

Petty, G.W., 1994b. Physical retrievals of over-ocean rain rate from multichannel microwave imagery. Part II: Algorithm implementation. *Meteorol. Atmos. Phys.,* **54,** 101–122.

Petty, G.W., 1997. An intercomparison of oceanic precipitation frequencies from 10 SSM/I rain-rate algorithms and shipboard present-weather reports. *J. Geophys. Res.,* **102,** 1757–1777.

Petty, G.W., and M.D. Conner, 1994. *Identification and classification of transient signatures in over-land SSM/I imagery.* 7th Conf. on Satellite Meteorology and Oceanography. 6–10 June, 1994, Monterey CA, American Meteorological Society, Boston, 193–196.

Petty, G.W., and K.B. Katsaros, 1990. Percipitation observed over the South China Sea by the Nimbus-7 scanning multichannel Microwave Radiometer during winter MONEX. *J. Applied Meteorol.,* **29,** 273–287.

PIP-II, precipitation algorithm intercomparison/validation project, 1995. *Workshop on precipitation retrieval algorithms.* Huntsville, AL., June 19–22, 1995. (Available from Dr. E. Smith, FSU, Dept. of Meteorology, Tallahassee, FL. 32306).

Pitter, R.L., and H.R. Pruppacher, 1974. A numerical investigation of collision efficiencies of simple ice plates colliding with supercooled water drops. *J. Atmos. Sci.,* **31,** 551–559.

Platt, C.M., S.A. Young, A.I. Carswell, S.R. Pal, M.P. McCormick, D.M. Winkler, M. DelGuasta, L. Stefanutti, W.L. Eberhard, M. Hardesty, P.H. Flamant, R. Valentin, B. Forgan, G.G. Gimmestad, H. Jäger, S.S. Khmelevtsov, I. Kolev, B. Kaprieolev, Da-ren Lu, K. Sassen, V.S. Shamanaev, O. Uchino, Y. Mizuno, U. Wandinger, C. Weitkamp, A. Ansmann and C. Woolridge, 1994. The experimental cloud lidar pilot study (ECLIPS) for cloud-radiation research. *Bull. Amer. Meteorol. Soc.,* **75,** 1635–1654.

Powell R.L., W.J. Hall, C.H. Hyink, Jr., L.L. Sparks, G.W. Burns, M.G. Scroger, and H.H. Plumb, 1975. *Thermocouple Reference Tables based on the IPTS–68, NBS Monograph-125.* OMEGA Engineering, Inc., Stamford, CT., 404 pp.

Price, J.C., 1984. Land surface temperature measurements from the split window channels of the NOAA–7 Advanced Very High Resolution Radiometer. *J. Geophys. Res.,* **89,** 7231-7237.

Pruppacher, H.R., and J.D. Klett, 1980. *Microphysics of Cloud and Precipitation,* Dordrecht, Holland/Boston, MA: D. Riedel Publishing Co., 714 pp.

Radiometrics Corp., 1995. *Radiometrics microwave water vapor radiometer.* Available from Radiometrics Corp., 3190 So. Wadsworth Blvd., Suite 100, Lakewood, CO 80227, 2 pp.

Radke, L.F., and D.A. Hegg, 1972. The shattering of saline droplets upon crystallization. *J. Rech. Atmos. (Atmos. Res.),* **6,** 447–455.

Radke, L.F., and P.V. Hobbs, 1991. Humidity and particle fields around some small cumulus clouds. *J. Atmos. Sci.,* **48,** 1190–1195.

Ralph, F.M., P.J. Neiman, D.W. van de Kamp and D.C. Law, 1995. Using spectral moment data from NOAA's 404 MHz radar wind profilers to observe precipitation. *Bull. Amer. Meteorol. Soc.,* **76,** 1717–1739.

Randel, D.L., T.H. Vonder Haar, M.A. Ringerud, G.L. Stephens, T.J. Greenwald, and C.L. Combs, 1996. A new global water vapor dataset. *Bull. Amer. Meteorol. Soc.,* **77,** 1233–1246.

Ranz, W.E., and W.R. Marshall, 1952. Evaporation from drops, Part II. *Chem. Eng. Progress,* **48,** 173–180.

Read, W.G., J.W. Waters, D.A. Flower, L. Froidevaux, R.F. Jarnot, D.L. Hartmann, R.S. Harwood, and R.B. Rood, 1995. Upper-tropospheric water vapor from UARS MLS. *Bull. Amer. Meteorol. Soc.,* **76,** 2381–2390.

REMS•Talk, 1995. A Newsletter for Remote Environmental Monitoring System Users, Sept. Available from Handar, 1288 Reamwood, Sunnyvale, CA 94089-2233, 16 pp.

Rider, N.E., 1957. Water loss from various land surfaces. *Q. J. Roy. Meteorol. Soc.,* **83,** 181–193.

Robertson, J.H., and D.I. Katz, 1995. Climatronics' novel sonic anemometer. Available from Climatronics Corp., 140 Wilbur Pl., Bohemia, NY 11716, 4 pp.

Rogers, R.R., and M.K. Yau, 1991. *A Short Course in Cloud Physics.* New York: Pergamon Press, 307 pp.

Rosenfeld, D., and A. Gagin, 1989. Factors governing the total rainfall yield from continental convective clouds. *J. Appl. Meteor.,* **28,** 1015–1030.

Rosenfeld, D., and W.L. Woodley, 1989. Effects of cloud seeding in West Texas. *J. Appl. Meteor.,* **28,** 1009–1014.

Salter, M. deC., 1921. *The rainfall of the British Isles.* London: University of London Press.

Sassen, K., 1975. Laser depolarization "bright band" from melting snowflakes. *Nature,* **255,** 316–318.

Sassen, K., 1977. Optical backscattering from near-spherical water, ice, and mixed phase drops, *Appl. Optics,* **16,** 1332–1341.

Sassen, K., 1978. Air-truth lidar polarization of orographic clouds. *J. Appl. Meteorol.,* **17,** 73.

Sassen, K. and B.S. Cho, 1992. Subvisual-thin cirrus lidar dataset for satellite verification and climatological research. *J. Appl. Meteorol.,* **32,** 1548–1558.

Saxena, V.K., 1996. Short bursts of cloud condensation nuclei (CCN) by dissipating clouds at Palmer Station, Antarctica. *Geophys. Res. Ltrs.,* **23,** 69–72.

Saxena, V.K., and G.F. Fisher, 1984. Water solubility of cloud active aerosols. *Aerosol Sci. and Technol.,* **3,** 335–344.

Saxena, V.K. and N.-H. Lin, 1990. Cloud chemistry measurements and estimates of acidic deposition on an above cloudbase coniferous forest. *Atmos. Environ.,* **24A,** 329–352.

Saxena, V.K., R.E. Stogner, A.H., Hendler, T.P. DeFelice, N.-H., Lin, and R.J.-Y. Yeh, 1989. Monitoring the chemical climate of the Mt. Mitchell State Park for evaluating its impact on forest decline. *Tellus,* **41B,** 92–109.

Schaefer, V.J., and R.J. Cheng, 1971. The production of ice crystal fragments by sublimation and electrification. *J. Rech. Atmos. (Atmos. Res.),* **5,** 5–10.

Shipley, S.T., D.H. Tracy, E.W. Eloranta, J.T. Trauger, J.T. Sroga, F.L. Roesler, and J.A. Weinman, 1983. High spectral resolution lidar to measure optical scattering properties of atmospheric aerosols. 1: Theory and instrumentation. *Appl. Optics,* **22,** 3716–3724.

Simpson, J., R.F. Adler, and J.R. North, 1988. A proposed Tropical Rainfall Measuring Mission (TRMM) satellite. *Bull. Amer. Meteorol. Soc.,* **69,** 278–295.

Smiley, V.N., B.M. Morley, and B.M. Whitcomb, 1976. Atmospheric investigations in polar regions using dye lasers. *Optics Communications,* **18,** 188–189.

Smiley, V.N., and B.M. Morley, 1979. Polarization measurements on Antarctic ice clouds by lidar. (In prep). Cited in Smiley, Whitcomb, Morley, and Warburton, 1980.

Smiley, V.N., B.M. Morley, and J.A. Warburton, 1979. Lidar determination of atmospheric ice crystal layers at South Pole during clear-sky precipitation (*Submitted to J. Applied Meteorol.*—cited in Smiley et al., 1980).

Smiley, V.N., B.M. Whitcomb, B. Morley and J.A. Warburton, 1980. Lidar determinations of atmospheric ice crystal layers at South Pole during clear-sky precipitation. *J. Appl. Meteorol.,* **19,** 1074–1091. (Note: Smiley, V.N., B. Morley, and J.A. Warburton, 1980. Lidar operations at Palmer Station. *Antarctic Journal,* 205–206 is also helpful, but not as easily found).

Smith, E.A., A. Mugnai, H.J. Cooper, G. Tripoli, and X. Xiang, 1992. Foundations for statistical-physical precipitation retrieval from passive microwave satellite measurements. Part I: Brightness temperature properties of a time-dependent cloud-radiation model. *J. Applied Meteorol.,* **31,** 506–531.

Smith, W.L., H.M. Woolf, H.B. Howell, H.-L. Huang, and H.E. Revercomb, 1987. The simultaneous retrieval of atmospheric temperature and water vapor profiles-application to measurements with the high spectral resolution interferometer sounder (HIS). In A. Deepak, H. Fleming, and J. Theon, eds., *Advances in Remote Sensing Retrieval Methods,* Hampton, VA: A. Deepak Publishing, 189–202.

Smith, W.L., R.O. Knuteson, H.E. Revercomb, W. Feltz, H.B. Howell, W.P. Menzel, N.R. Nalli, O. Brown, J. Brown, P. Minnett, and W. McKeown, 1996. Observations of the infrared radiative properties of the ocean: Implications for the measurement of sea surface temperature via satellite remote sensing. *Bul. Amer. Meteorol. Soc.,* **77,** 41–52.

Snider, J.B., 1971. Recent results in ground-based sensing of atmospheric temperature profiles. Proceedings of Symposium on Air Pollution, Turbulence and Diffusion, 7–10 December, 56–65.

Snider, J.B., and E.R. Westwater, 1972. Radiometry. Chapter 15 in *Remote sensing of the troposphere,* Ed. V.E. Derr. U.S. Dept. of Commerce, NOAA, Wave Propagation Lab, ERL-Electrical Engineering Dept., Colorado University, Boulder, CO., 33 pp.

Spencer, R.W., and J.R. Christy, 1992. Precision and radiometric validation of satellite gridpoint temperature anomalies, Part I. MSU channel 2. *J. Climate,* **5,** 847–857.

Spencer, R.W., H.H. Goodman, and R.E. Hood, 1989. Precipitation retrieval over land and ocean with SSM/I- Identification and characteristics of the scattering signal. *J. Atmos. Oceanic Technol.,* **6,** 254–273.

Spencer, R.W., F.J. LaFontaine, T.P. DeFelice and F.J. Wentz, 1998. Tropical Oceanic Precipitation Changes after the 1991 Pinatubo Eruption. *J. Atmos. Sci.,* In Press.

Spencer, R.W., R.E. Hood, F.J. LaFontaine, E.A. Smith, R. Platt, J. Galliano,V.L. Griffen, and E. Lobl, 1994. High-resolution imaging of rain systems with the advanced microwave precipitation radiometer. *J. Atmosph. Ocean. Technol.,* **11,** 849–857.

Spinhirne, J.D., J.A.R. Rall and V.S. Scott, 1996. Compact Eye Safe Lidar Systems. *The Review Laser Engineering,* **23,** 112–118.

Spinhirne, J.D., W.D. Hart and D.L. Hlavka, 1996b. Cirrus infrared parameters and shortwave reflectance relations from observations. *J. Atmos. Sci.,* **53,** 1438–1458.

Sroga, J.T., E.W. Eloranta, J.T. Trauger, S.T. Shipley, F.L. Roesler and P.J. Tryon, 1983. High spectral resolution lidar to measure optical scattering properties of atmospheric aerosols. 1: Calibration and data analysis. *Appl. Optics,* **22(23),** 3725–3732.

Staelin, D.H., K.F. Kunzi, R.L. Pettyjohn, K.L. Poon, R.W. Wilcox and J.W. Waters, 1976. Remote sensing of atmospheric water vapor and liquid water with the Nimbus 5 Microwave spectrometer. *J. Appl. Meteorol.,* **15,** 1204–1214.

Stearns, C.R., and G. Wendler, 1988. Research results from Antarctic automatic weather stations. *Rev. Geophys.,* **26,** 45–61.

Stearns, C.R., J.T. Young and B. Sinkula, 1995. Two years of IR images south of 40 deg S every three hours. *Proc. Fourth Conf. on Polar Meteorology and Oceanography,* Dallas, TX, Amer. Meteorol. Soc., **112.**

Stearns, C.R., 1996. *Antarctic Automated Weather Station Program.* (Available from Prof. Dr. C. Stearns, Univ. of Madison, Madison, WI).

Stephens, G.L., D.L. Jackson and J.J. Bates, 1994. A comparison of SSM/I and TOVS column water vapor data over global oceans. *Meteor. Atmos. Phys.,* **54,** 183–201.

Stewart, R.E., 1992. Precipitation types in the transition region of winter storms. *Bull. Amer. Meteorol. Soc.,* **73,** 287–296.

Sutton, O.G., 1977. *Micrometeorology.* New York: R.E. Krieger Publ. Co., 333 pp.

Tanner, B.D., 1990. Automated weather stations. In N.G. Goel and J.M. Norman, eds., Instrumentation for studying vegetation canopies for remote sensing in optical and thermal infrared regions. *Remote Sensing Reviews,* **5,** 73–98.

Tanner, C.B., 1960. Energy balance approach to evapotranspiration from crops. *Soil Sci. Soc. Am. Proc.,* **24,** 1–9.

Thornthwaite, C.W., and B. Holzman, 1939. The determination of evaporation from land and water surfaces. *Mon. Wea. Rev.,* **67,** 4–11.

Thornthwaite, C.W., and J.R. Mather, 1954. The computation of soil moisture in estimating soil tractionability from climatic data. In *Publ. in Climat.,* **7,** 397–402.

Thornthwaite, C.W., and J.R. Mather, 1955. The water balance. *In Publ. in Climat.,* **8.**

Tremblay, A., and H. Leighton, 1986. A three-dimensional cloud chemistry model. *J. Clim. Appl. Meteorol.,* **18,** 652–671.

Turner, J., D. Bromwich, S. Colwell, S. Dixon, T. Gibson, T. Hart, G. Heinemann, H. Hutchinson, K. Jacka, S. Leonard, M. Lieder, L. Marsh, S. Pendlebury, H. Phillpot, M. Pook, and I. Simmonds, 1996. The Antarctic first regional observing study of the troposphere (FROST) project. *Bull. Amer. Meteorol. Soc.,* **77,** 2007–2032.

Twomey, S., 1955. The composition of hygroscopic particles in the atmosphere. *J. Meteorol.,* **11,** 334–338.

Twomey, S., and K.N. McMaster, 1955. The production of condensation nuclei by crystallizing salt particles. *Tellus,* **7,** 458–461.

Twomey, S., M. Piepgrass, and T.L. Wolfe, 1984. An assessment of the impact of pollution on global cloud albedo. *Tellus,* **36B,** 356–366.

U.S. Department of Commerce, National Weather Bureau (Service), 1963. *Manual of Barometry* (WBAN), Vol. I, First edition. Available from U.S. Govt. Printing Office, Washington, D.C.

U.S. EPA, 1976. *Quality Assurance Handbook for Air Pollution Measurement Systems: Vol. I, Principles.* **EPA–600/9–76–005**. Environmental Monitoring Systems Laboratory, Environmental Protection Agency, Research Triangle Park, NC, 27711.

U.S. EPA, 1980. *Interim Guidelines and Specifications for Preparing Quality Assurance Project Plans.* Office of Monitoring Systems and Quality Assurance, Environmental Protection Agency, **QAMS–005/80,** Washington, D.C., 20460.

Uttal, T., J.B. Snider, R.A. Kropfli, and B.W. Orr, 1990. A remote sensing method of measuring atmospheric water vapor fluxes: Application to winter mountain storms. *J. Appl. Meteorol.,* **29,** 22–33.

Vaisala, 1989. RS–80 Radiosondes. Upper-air systems product information. Reference No. R0422–2. (Available from Vaisala Inc., 100 Commerce Way, Woburn, MA 01801–1068.) 5 pp.

Vaisala, 1994. *RS80 Radiosondes.* Tech. Brief. Vaisala, Inc. (Available from Vaisala Inc., 100 Commerce Way, Woburn, MA 01801–1068.) 3 pp.

Vaisala, 1995a. GPS wind finding information release. Vaisala, Inc. Techn. Report R599en 1995–03. (Available from Vaisala Inc., 100 Commerce Way, Woburn, MA 01801–1068.) 4 pp.

Vaisala, 1995b. *High-Precision Radiosonde RS90.* Tech. Brief. Vaisala, Inc. (Available from Vaisala Inc., 100 Commerce Way, Woburn, MA 01801–1068.) 3 pp.

Vaisala, 1997. The future of windfinding and new dropsondes. *Vaisala News* **142,** 5–20. (Available from Vaisala Inc., 100 Commerce Way, Woburn, MA 01801–1068.)

Vivekanandan, J., J. Turk, and V.N. Bringi, 1993. Comparisons of precipitation measurements by the advanced microwave precipitation radiometer and multiparameter radar. *IEEE Trans. Geosci. & Remote Sensing,* **31,** 860–870.

Vong, R.J., and A.S. Kowalski, 1994. Eddy correlation measurements of size-dependent cloud droplet turbulent fluxes to complex terrain. *Tellus,* **47B,** 331–352.

Wall, C.N., R.B. Levine, and F.E. Christensen, 1972. *Physics Laboratory Manual,* 3rd ed. Englewood Cliffs, NJ: Prentice Hall, 419 pp.

Wallace, J.M., and P.V. Hobbs, 1977. *Atmospheric Science-An introductory Survey.* New York: Academic Press, 467 pp.

Wang, J. Y., 1975. *Instruments for Physical Environmental Measurements.* Milieu Information Service, 276 pp.

Wang, T.-I., and J.D. Crosby, 1993. Taking rain gauges to sea. *Sea Technology,* **34,** 61–65.

Warburton, J.A., 1969. Trace silver detection in precipitation by atomic absorption spectrophotometry. *J. Appl. Meteorol.,* **8,** 464–466.

Warburton, J.A., and T.P. DeFelice, 1986. Oxygen isotopic composition of central Sierra Nevada precipitation. Part I. Identification of ice phase water capture regions in winter storms. *Atmospheric Research,* **20,** 11–22.

Warburton, J.A., B.B. Demoz, and R.H. Stone, 1993. Oxygen isotopic variations of snowfall from winter storms in the central Sierra Nevada; relation to ice growth microphysics and mesoscale structure. *Atmospheric Research,* **29,** 135–151.

Warburton, J.A., J.V. Molenar, M.S. Owens and A. Anderson, 1980. Heavy metal enrichment in Antarctic precipitation and near surface snow. *Pageoph. (Birkhäuser Verlag, Basel),* **118,** 1–5.

Ward, R.C., 1967. *Principles of Hydrology.* New York: McGraw-Hill Publ. Co., 403 pp.

Wentz, F.J., 1988. *SSMI/I $T_a$ Tape User's Manual.* (Available from Remote Sensing Systems, Inc., 1101 College Ave., Suite 220, Santa Rose, CA 95404.) 36 pp.

Wentz, F.J., 1991. *SSM/I Antenna Temperature Tape User's Manual, Revision 1.* (Available from Remote Sensing Systems, Inc., 1101 College Ave., Suite 220, Santa Rose, CA 95404.) 72 pp.

Wentz, F.J., 1992. Measurement of oceanic wind vector using satellite microwave radiometers. *IEEE Trans. Geosci. Remote Sens.,* **30,** 960–972.

Wentz, F.J., and R.W. Spencer, 1996. Oceanic rainfall retrievals from a unified ocean algorithm. *J. Atmos. Sci.* In Press.

Westwater, E.R., 1972. Ground-based determination of low altitude temperature profiles by microwaves. *Monthly Weather Rev.,* **100,** 15–28.

Wexler, 1970. Measurement of humidity in the free atmosphere near the surface of the earth. *Meteorological Observations and Instrumentation. Meteorol. Monograph,* **33,** 262–282. American Meteorological Society, Boston, MA. 455 pp.

Whiteman, C.D., and X. Bian, 1996. Solar semidiurnal tides in the troposphere: Detection by radar profilers. *Bull. Amer. Meteorol. Soc.,* **77,** 529–542.

Whiteman, D.N., S.H. Melfi, and R.A. Ferrare, 1992. Raman lidar system for the measurement of water vapor and aerosols in the Earth's atmosphere. *Appl. Opt.,* **31,** 3068–3082.

Whitlow, S., P.A. Mayewski, and J.E. Dibb, 1992. A comparison of major chemical species seasonal concentration and accumulation at the south pole and summit, Greenland. *Atmos. Environ.,* **26A,** 2045–2054.

Wielicki, B.A., B.R. Barkstrom, E.F. Harrison, R.B. Lee III, G.L. Smith, and J.E. Cooper, 1996. Clouds and the earth's radiant energy system (CERES): An earth observing system experiment. *Bul. Amer. Meteorol. Soc.,* **77,** 853–868.

Wieringa, J., 1967. Evaluation and design of wind vanes. *J. Appl. Meteorol.,* **6,** 1114–1122.

Wilheit, T.T., A.T.C. Chang, and L. Chiu, 1991. Retrieval of monthly rainfall indices from microwave radiometric measurements using probability distribution function. *J. Atmos. Oceanic Technol.,* **8,** 118–136.

Winters, W., A. Hogan, V. Mohnen, and S. Barnard, 1979. ASRC Airborne cloudwater collection system. *ASRC SUNYA Publ.* **#728,** May, 50 pp.

Wood, R.C., 1963. The infrared hygrometer: Its application to difficult humidity measurement problems. In *Humidity and Moisture,* **1,** 492–504, New York: Reinhold, 687 pp.

Wu, X., J.J. Bates, and S.J.S. Khalsa, 1993. A climatology of the water vapor band brightness temperatures from NOAA operational satellites. *J. Climate,* **6,** 1282–1300.

Wurman, J., 1994. Vector winds from a single-transmitter bistatic dual-doppler radar network. *Bull. Amer. Meteorol. Soc.,* **75,** 983–994.

Wurzler, S., P. Respondek, A.I. Flossmann, and H.R. Pruppacher, 1994. Simulation of the dynamics, microstructure and cloud chemistry of a precipitating and a non-precipitating cloud by means of a detailed 2-D cloud model. *Beitr. Phys. Atmosph.,* **67,** 313–319.

Yuan, S.W., 1967. *Foundations of Fluid Mechanics,* Englewood Cliffs, NJ: Prentice Hall, 608 pp.

Young, K.C., 1993. *Microphysical Processes in Clouds.* New York: Oxford Univ. Press, 427 pp.

APPENDIX A

# Basic Electronics of Measuring Devices

## A.1 Basic DC Circuits

Most meteorological measuring devices contain electrical components that convert changes in a physical quantity into changes in voltage potential, electrical current, electrical resistance, or frequency. This section provides the fundamentals for describing the aforementioned conversion using *direct current (dc) circuits*.

The voltage drop (V), or electric potential (E), across a resistance (R) of a conductor, when current (I) passes through the resistance will, if the conductor is ohmic, have the following relationship known as Ohm's law:

$$V = IR$$

or, (A.1)

$$V = A\Omega = \{[\text{COULOMB (charge)}]s^{-1}\}\text{OHMS}$$

where the units of I are amperes, A, and the units of resistance, R, are Ohms, Ω. Equation A.1 holds as long as the temperature, composition, and shape of the conductor remain constant, since the resistance of the conductor depends on these three

---

**Example A.1**

Find the current if a 1.5 v voltage drop is measured across a 150 Ω resistor. $V = IR \Rightarrow I = 1.5V/150\,\Omega = 0.01$ Amps = 10 milliamperes or 10 mA.

---

quantities. A conductor that does not obey Equation A.1 is non-ohmic. If the circuit yields a signal that varies with time, i.e., an *alternating current circuit*, ac, then the resistor is replaced by a capacitor, and current is related to the charge on the capacitor as will be seen in section 4 below.

Figure A.1 illustrates the conventional flow of current, namely "+" to "−", and electrons, i.e., "−" to "+" using a simple electronic circuit. A simple electronic circuit contains a voltage source and one electrical component (i.e., resistor, capacitor, inductor, diode, etc.). The electrical components may be arranged in series with, or parallel to, a voltage source. For example, let the electrical components be resistors.

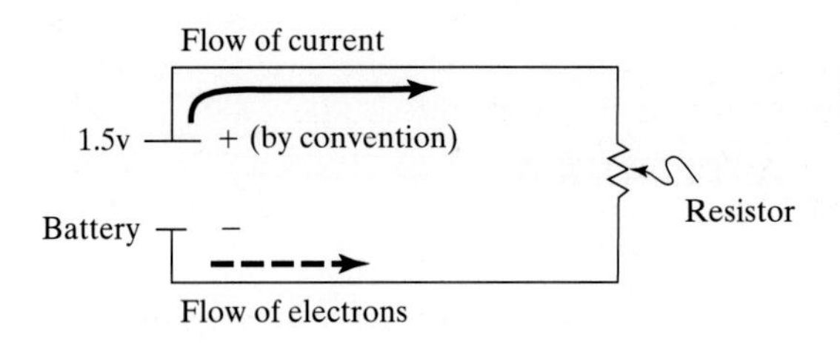


**Fig. A.1.** *Conventional flow of current in a simple circuit.*

(a)

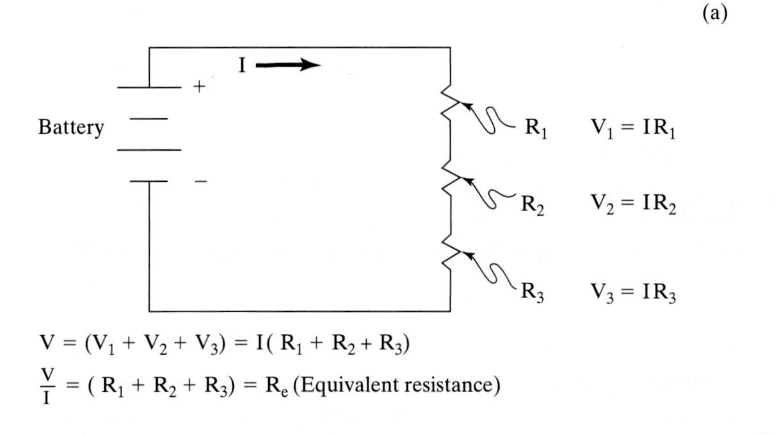


$$V = (V_1 + V_2 + V_3) = I(R_1 + R_2 + R_3)$$

$$\frac{V}{I} = (R_1 + R_2 + R_3) = R_e \text{ (Equivalent resistance)}$$

(b)

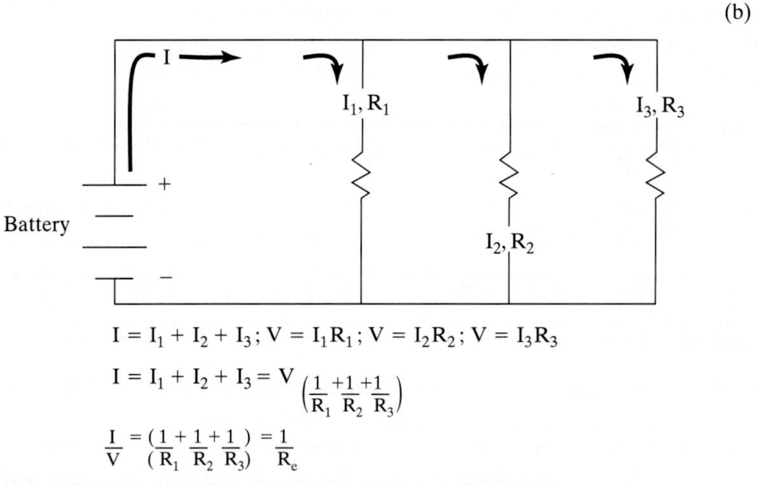


$$I = I_1 + I_2 + I_3;\ V = I_1R_1;\ V = I_2R_2;\ V = I_3R_3$$

$$I = I_1 + I_2 + I_3 = V\left(\frac{1}{R_1} + \frac{1}{R_2} + \frac{1}{R_3}\right)$$

$$\frac{I}{V} = \left(\frac{1}{R_1} + \frac{1}{R_2} + \frac{1}{R_3}\right) = \frac{1}{R_e}$$

**Fig. A.2.** *Example of a (a) series and (b) parallel circuit.*

Resistors that are configured along the current flow path within a given circuit loop make up a series circuit. The current is the same in all parts of a series circuit (Fig. A.2a). Resistors that branch into more than one path make up a parallel circuit. The voltage is constant through a parallel circuit (Fig. A.2b). It is clear from Fig.

A.2b that the equivalent resistance of a parallel circuit, $R_e$, parallel, is less than the smallest value resistor in the circuit. Since there is a voltage drop across a resistor, and a voltage drop is directly related to power loss or *dissipation* (i.e., kind of like the result from plugging the air conditioning, stereo, TV, and stove into the same outlet that was only meant to handle the stereo). The power dissipated as current, I, passes through a resistor with resistance, R, may be given as,

$$P = IV = I^2R = V^2R^{-1} \tag{A.2}$$

where V is defined as the voltage drop across a resistor.

## A.2 *Complicated Circuits*

Circuits are not always as simple as those above, and their configuration depends on the task for which they are designed. This section provides the basics of how to determine the currents and voltage drops, etc. in more complicated circuits. The determination of current and voltage drops in more complicated circuits are generally summarized by what is known as Kirchoff's laws. Kirchoff's laws state that

1. The sum of the voltage drops around a loop equals zero (voltage law), and
2. The sum of the currents flowing into and away from a junction equals zero (current law).

The voltage law yields the following expressions for the circuit in Fig. A.3: (left hand loop): $V - (I_1 R_1) = 0$; (right hand loop): $V - (I_2 R_2) = 0$. The current law yields the following expressions for the circuit in Fig. A.3: (junction A): $I - I_1 - I_2 = 0$; (junction B): $I_1 + I_2 - I = 0$. (Note: Think "into point equals out of point," then do the algebra).

The circuits can become more complicated than in Fig. A.3, and one only needs to apply the formulations for parallel and series resistor circuits until a simplified circuit is obtained and its resistance (or equivalent resistance), easily determined. That is, put any resistors in series together first, then put parallel ones together, then series, then parallel, etc., until you achieve a circuit that is similar to

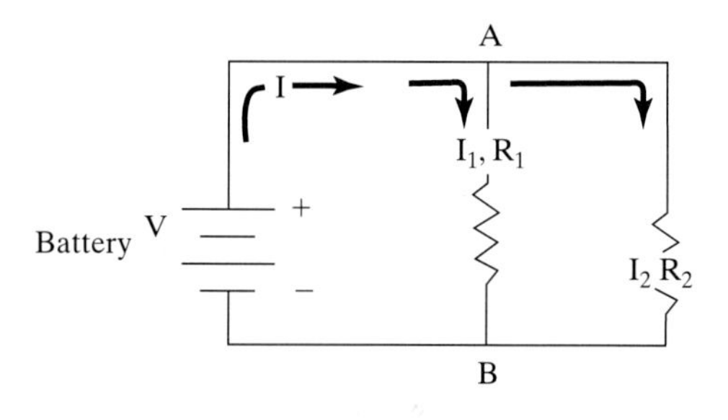


**Fig. A.3.** *Circuit to illustrate Kirchoff's law.*

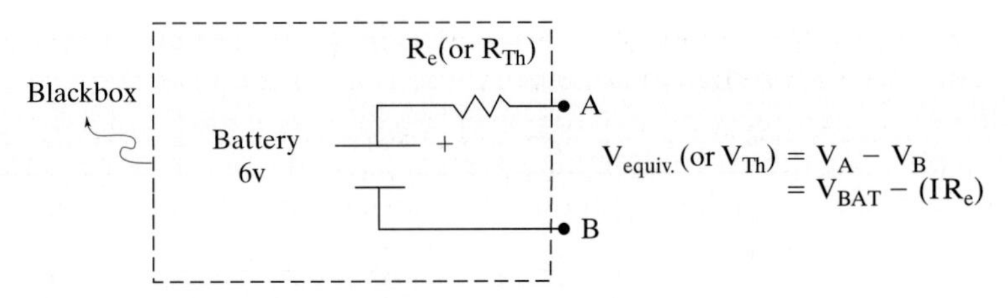


**Fig. A.4.** *Conceptual circuit for Thevenin's Theorem.*

the one in Fig. A.1. Let's call the latter, the "brute force" method, just for the sake of discussion. The "brute force" method can get pretty tedious, to say the least, and probably helped provide the impetus for Thevenin to come up with a slightly quicker and less tedious approach.

Thevenin simply treats a component of interest as a black box in which there exists some representative resistance, and one measures the voltage drop across it. The latter has proven to be useful under essentially all conditions, and is the basic premise for what is known as *Thevenin's theorem.* Thevenin's theorem may be summarized in four steps:

1. Remove the component of interest—label leads A and B.
2. The Thevenin equivalent voltage, $V_{Th}$, equals the voltage between terminals A and B (Fig. A.4).
3. Short the voltage sources, and calculate the equivalent Thevenin resistance, $R_{Th}$, from A to B.
4. Replace the component (from between A and B) in the equivalent circuit.

## A.3 *Basic Measuring Instruments with DC Circuits*

There are three fundamental DC measuring instruments, namely, the Galvanometer, the potentiometer, and the Wheatstone bridge. They follow.

***Galvanometer.*** The galvanometer is primarily a current measuring device, but it can be configured to measure voltage and resistance. The galvanometer has some internal resistance caused from a coiled wire that is placed within a magnetic field, and by or from a meter. The use of the galvanometer to measure current, I, requires that the measured current remains below the full scale current, $I_{fs}$. To ensure this, the current is shunted before it reaches the meter (Fig. A.5). Consequently, the scale (current output) on the meter is modified by changing the shunt resistor.

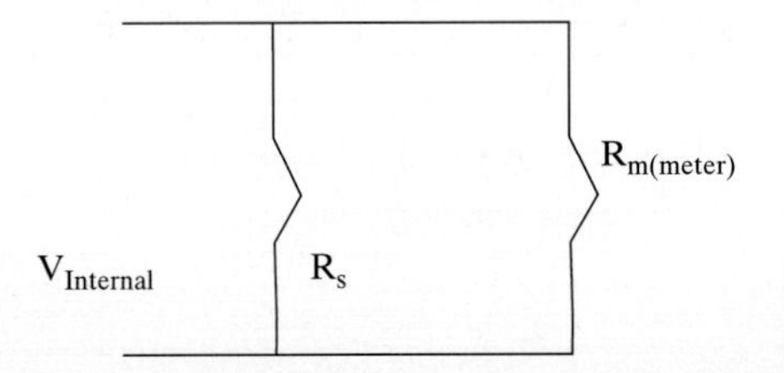


**Fig. A.5.** *Definition of a shunting resistor,* $R_s$.

---

**Example A.2.**

What is the equivalent resistance, and the current through the scale on the meter of the above circuit (Find $I_{scale}$)? Let $I_s$ be the current through the shunt, and $I_m$ be the current through the meter.

$$R_{equiv.} = (R_s R_m)(R_s + R_m)^{-1}$$
$$V_{internal} = I_{scale} R_{equiv.} \tag{A.3}$$
$$I_s = V_{Internal}/R_s \;\; I_m = V_{Internal}/R_m \Rightarrow I_s R_s = I_m R_m$$
$$I_{scale} = I_s + I_m \Rightarrow R_s = R_m[I_m(I_{scale} - I_m)^{-1}]$$

Note: Ohm/voltage rating: $\Omega$/V rating = $(I_{fullscale})^{-1} \approx 20$ k$\Omega$/Volts (a typical value).

---

The galvanometer, G, may be used as a voltmeter, or ohmmeter by simply connecting it in series with another meter, or as shown in Fig. A.6, respectively.

***Potentiometer.*** The potentiometer is essentially a precision voltage divider. The current, I, and $V_{out}$ in Fig. A.7 may be expressed in terms of the resistance of the load, $R_L$, and $R_e$ as

$$I = V_{equiv.}(R_e + R_L)^{-1}; V_{out} = (V_{equiv.} R_L)(R_e + R_L)^{-1}. \tag{A.4}$$

Note that $R_e$ and $V_{equiv.}$ are essentially the Thevenin resistance and voltage respectively. The standard potentiometer has two modes of operation, namely calibration and measurement (Fig. A.8). One can use Thevenin's theorem to find the fraction, $\alpha$, of the adjustable resistor in Fig. A.8 required to find the unknown voltage when the potentiometer is placed in measurement mode. The potentiometer is extremely accurate and

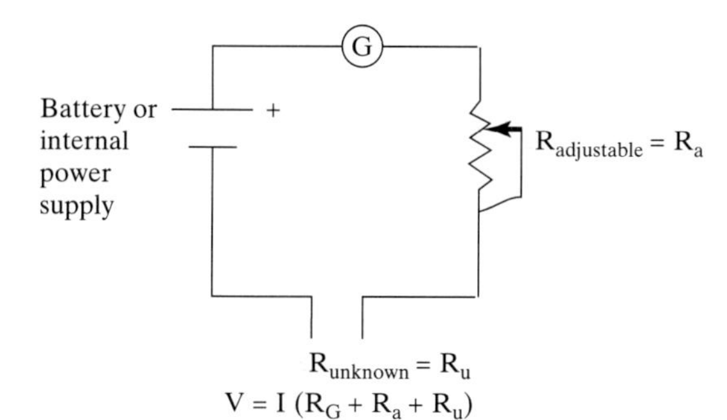


**Fig. A.6.** *A galvanometer configured as an ohmmeter.*

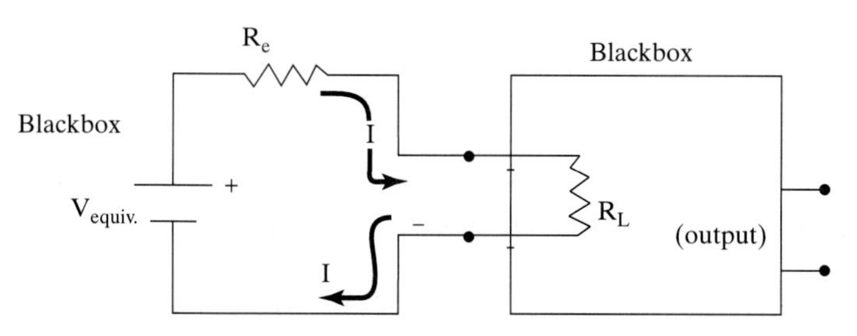


**Fig. A.7.** *The configuration of a voltage divider with a load.*

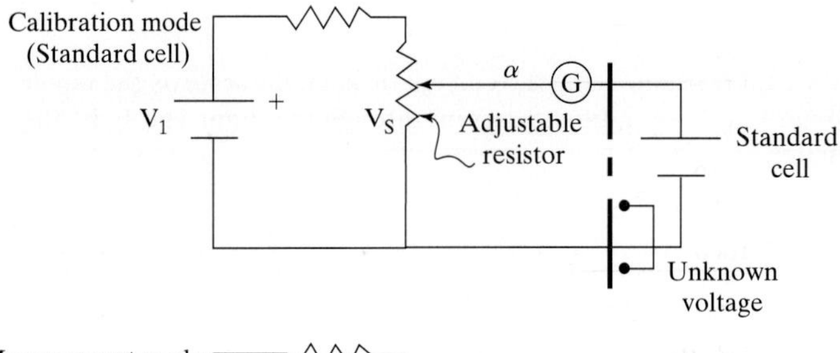


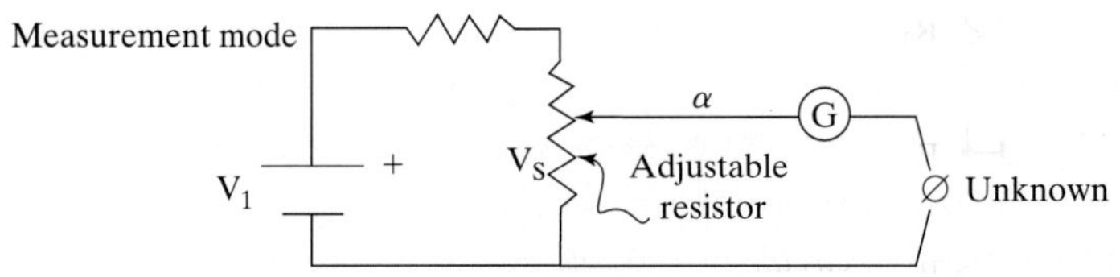


**Fig. A.8.** *The configuration of a standard potentiometer in calibration and measurement modes.*

has been automated. The automated version is known as the self-balancing potentiometer, and is exemplified by a high precision strip chart recorder.

The connection of a load to the potentiometer should not result in a power loss or dissipation, and you are in mega trouble if it does. However, knowing the size of a load that renders a device useless can come in handy. Consequently, suppose a load is connected to a potentiometer. What effect will it have on our potentiometer measurement? We can express the power dissipation due to the load, $P_L$, in terms of $R_L$, yielding

$$P_L = V_{out} I = [(V_{equiv.})^2 R_L](R_e + R_L)^{-2}. \tag{A.5}$$

The maximum power dissipation, $dP_L / dR_L$, may be readily obtained by essentially finding the maximum of a curve relating power and resistance, and is left to the reader. But what does the differential of Equation A.5 with respect to $R_L$ really tell us? Simply, the load puts a strain on the measuring device that could potentially cause the device to "report an erroneous value." When $R_L$ is 1000 times less than that of the measuring device, then the "strain" (power loss) is minimal (Table A.1). Maximum power loss occurs when the resistance of the load equals that of the measuring device, and is also known as the impedance matching condition. Given that $R_e$ and $R_L$ are related according to the following expression, $R_e = Ó R_L$ where Ó is a proportionality constant, then the voltage expression in Equation A.4 may be rewritten as

$$\frac{V_{out}}{V_{equiv.}} = \frac{Ó}{Ó + 1}. \tag{A.6}$$

The power dissipation is related to the difference between the output or load voltage and the equivalent voltage as shown in Table A.1.

***Wheatstone Bridge.*** The wheatstone bridge is the basis for most of the circuits used to measure resistance. All wheatstone bridge designs include a set of standard resistances, a signal source, and an output indicator. The particular design depends on whether the bridge will be used in a null-balanced or un-balanced configuration.

**Table A.1**

| Ó | %error(1−{$V_L/V_{equiv.}$}) |
|---|---|
| 0.1 | 90.9% |
| 1.0 | 50.0 |
| 10 | 9.09 |
| 100 | 0.99 |
| 1000 | 0.10 |

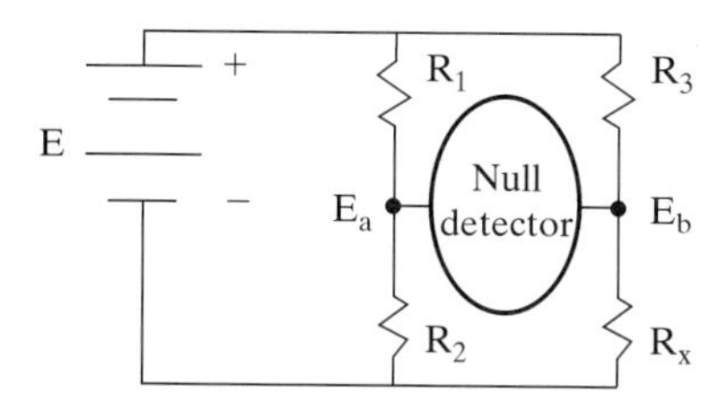


**Fig. A.9.** *Basic bridge circuit known as the null-balance wheatstone bridge.*

The sensitivity in the response of a bridge circuit increases as a balanced condition is approached.

The null-balance wheatstone bridge (Fig. A.9) is used to determine unknown resistances and temperature (using temperature sensitive resistors). It is widely used in AC and DC bridges, and many commercial models are available. One of the resistors, usually $R_2$, is a variable resistor, and $R_x$ is the unknown resistance. The bridge can be considered as two parallel voltage dividers. Given that $R_1$ and $R_3$ are equal in magnitude, the voltage drop across $R_x$ is compared with the voltage drop across $R_2$. Note, if the voltage drops are equal, the potential difference between the midpoints of the bridge is zero, and no current will flow through the null-detector (typically a galvanometer) when its switch is closed. Thus, the resistance of the unknown equals the resistance of $R_2$ (i.e., null-balance state). Given that $R_x \neq R_2$, current will flow through the galvanometer and $R_2$ is adjusted, either increased or decreased, until a null-balance is achieved. At balance:

$$R_2(R_1 + R_2)^{-1} = R_x(R_x + R_3)^{-1} \Rightarrow R_x = R_2(R_3/R_1). \quad \text{(A.7)}$$

The un-balanced bridge is preferred for use with automatic data logging systems capable of measuring voltages. The un-balance can be detected as a potential difference or as a current between the center points of the bridge.

## A.4 *Basic AC Circuits*

Alternating current (AC) circuits are generally similar to DC circuits, except for the voltage supply or power source to the circuit. Recall that the voltage in a DC circuit is constant. This is not the case in an AC circuit, where the voltage varies in time. The most common alternating current voltage, $V_{AC}$, variation is given by

$$V_{AC} = V_0 \sin(w\, t), \quad \text{(A.8)}$$

where $V_{AC}$ is the voltage applied to an alternating current circuit, $V_0$ is the amplitude of the wave (volts), w is the angular frequency (radians time$^{-1}$), and t is time. Similarly, the current in an AC circuit, $I_{AC}$, may be expressed as

$$I_{AC} = I_0 \sin(w' t), \tag{A.9}$$

where $I_0$ is the amplitude of the wave (amps) with an angular frequency of w′ (radians time$^{-1}$).

AC circuits are either resistive, inductive, or capacitive in nature, or some combination thereof. The simplest AC circuit is the resistive circuit, and is configured the same way as in Fig. A.1, except that an AC power source replaces the battery. Equations A.1 and A.2 are applicable, since the voltage and current are in phase with each other as they are in a DC circuit. Voltage and current are in phase with each other when their minima and maxima values occur at the same time. The inductive AC circuit is more complicated, than the former AC circuit, and is used in motors, generators, and transformers. Inductance (usually denoted as L) is due to a counter electromotive force that is produced when a magnetic field develops around a coil of wire. This counter electromotive force opposes a change in the current of the circuit, and, as a result, voltage "leads" the current. That is, the voltage maximum with respect to time occurs before the current maximum with respect to time in an inductive AC circuit. The difference in the aforementioned voltage and current waveforms is called the phase angle, and has a maximum of 90° when there is no resistor. Since most circuits do have some resistance (i.e., resistive-inductive circuit), the phase angle is actually less. An alternating current circuit that contains a capacitor is simply termed a capacitive alternating current circuit. Capacitors store electrical charge, q, in an amount proportional to the voltage applied (or potential difference), $V_{AC}$, i.e.,

$$q = CV_{AC} \tag{A.10}$$

where C is the capacitance of the capacitor (farads). If a capacitor is in a DC circuit then it will contain the same voltage as the DC power source when it's fully charged. The capacitor stops the flow of direct current in a DC circuit. In contrast, applying AC to a capacitor will cause the capacitor to charge and discharge, since the value of current changes. The most rapid change in voltage occurs at 0° and 180° positions as polarity changes from negative to positive. The positions of most rapid voltage change are also positions of maximum current development. The current, $I_{AC}$, is equal to the time rate of change in charge

$$I_{AC} = dq/dt = C dV_{AC}/dt( + V_{AC} dC/dt). \tag{A.11}$$

The latter bracketed term is generally considered to be zero, and will also be zero in our case. The current leads the voltage by a phase angle between 0° and 90°, with the latter angle associated with the purely capacitive circuit. There is no net power converted in the AC capacitive circuit. Capacitive circuits may be used for atmospheric moisture, pressure and temperature measurements. Capacitors placed in parallel are treated as resistors placed in series, whereas capacitors placed in series are treated as resistors placed in parallel (Fig. A.10).

(a)

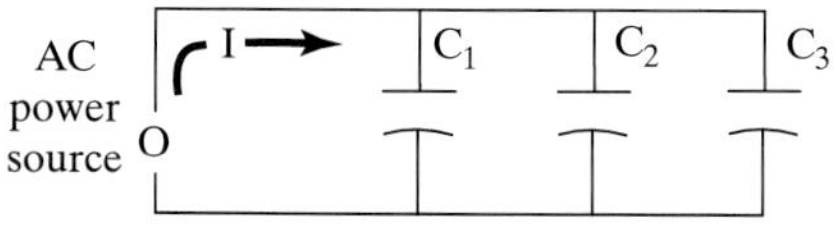

AC power source O

I → $C_1$ $C_2$ $C_3$

$$q = \Sigma^{3}_{i=1} q_i = V_{AC}(C_1 + C_2 + C_3);(V_{AC}/q) = (C_1 + C_2 + C_3)$$
$$= C_p \text{ (parallel capacitance)} \quad \text{(A.12)}$$

(b)

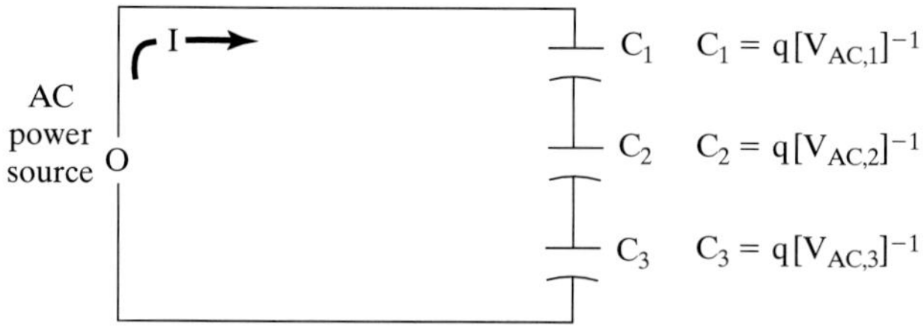

AC power source O

I →

$C_1$ $C_1 = q[V_{AC,1}]^{-1}$

$C_2$ $C_2 = q[V_{AC,2}]^{-1}$

$C_3$ $C_3 = q[V_{AC,3}]^{-1}$

$$V_{AC} = V_{AC,1} + V_{AC,2} + V_{AC,3} = q\left(\frac{1}{C_1} + \frac{1}{C_2} + \frac{1}{C_3}\right) = q\frac{1}{C_{s,e}} \quad \text{(A.13)}$$

$\frac{q}{V_{AC}} = C_{s,e}$, where $C_{s,e}$ is the equivalent capacitance for the capacitors in series

(c)

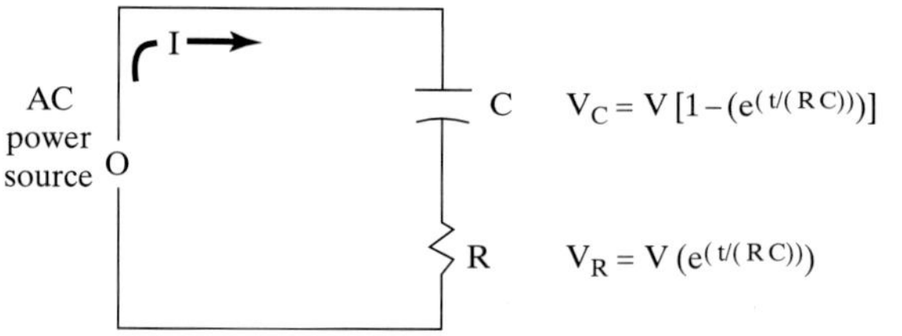

AC power source O

I →

C $V_C = V[1-(e^{(t/(RC))})]$ (A.14)

R $V_R = V(e^{(t/(RC))})$

$$V = V_c + V_R = (qC^{-1}) + (IR) \Rightarrow [q(RC^{-1})] + (dq/dt) = (VR^{-1}) \quad \text{(A.15)}$$

**Fig. A.10.** *An example of a circuit with capacitors placed in (a) parallel and (b) series. (c) An example of a circuit with a capacitor placed in series with a resistor, R(i.e., an RC circuit).* $V_c$ *is the voltage drop across the capacitor C, and* $V_R$ *is the voltage drop across the resistor R. R C is the time constant, or time (ohm coulombs/volt = s) when* $V_c = 0.632\,V$*, and* $V_c = V_R$ *at 5(0.693 R C).*

Impedance, Z, is the alternating current analogue of resistance, and is expressed as $Z=[(\text{resistance, } R)^2 - (\text{reactance, } X)^2]^{0.5}$. Thus if the AC circuit only had a resistor then Z = R at every frequency, f, or wavelength. The reactance is also related to a capacitance or an inductance present in the circuit. The AC circuit with a capacitor has a capacitance reactance of $X_{capacitor} = (2\pi f C)^{-1}$ and an inductive AC circuit with a capacitor has a inductance reactance of $X_{capacitor} = (2\pi f L)$, where L is the inductance (Henries).

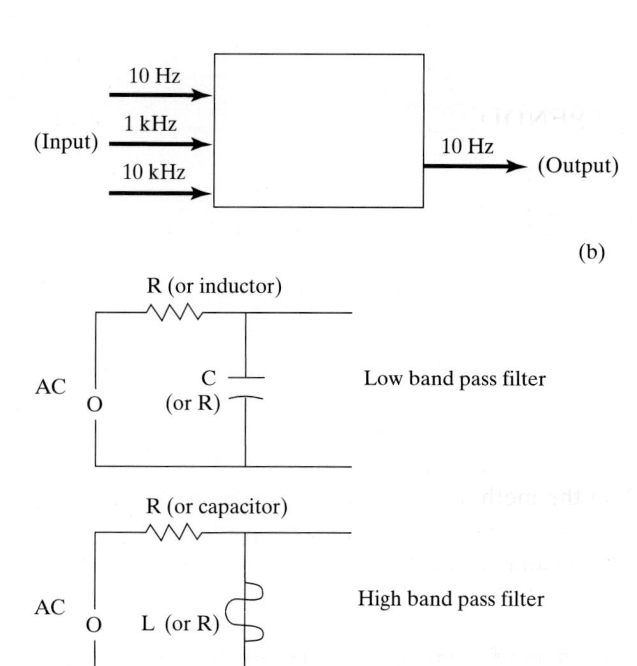


**Fig. A.11.** *Filter circuits. (a) Conceptual (high shown) and (b) Examples of low-and-high band bass filters using AC circuit configurations. For the low band pass filter example, AC frequency is sent through a resistor and then a capacitor, or an inductor and then a resistor, then only the low frequency bands are passed. The reverse of the latter holds when high frequency bands are passed.*

There are three types of filter circuits used in AC devices to separate one range of frequencies from another, namely, low-pass filters, high-pass filters, and band-pass filters. The low-pass filters pass low frequencies (long wavelengths) and block high frequencies (short wavelengths), the high-pass filters pass high frequencies and block lower frequencies (Fig. A.11a), and the band-pass filters pass a mid-range of frequencies and blocks higher and lower frequencies. Filter circuits typically have a resistance and inductance or capacitance (Fig. A.11b). Instruments, such as lidars or radiometers, that measure particular portions of the electromagnetic radiation might use one of these filters.

APPENDIX B

# Analog to Digital Conversion Basics

Analog electronics are based on the method of determining one quantity in terms of another, and are used in measuring devices designed for measuring continuously changing values. A simple example of an analog device is the mercury-in-glass thermometer. Digital electronics is considered a counting operation, wherein, the digital device typically counts the number of generated voltage pulses. The counts are then used to perform some function, such as counting or integrating. Figure B.1 shows some of the relevant theory. **Gates** (see glossary) determine whether or not frequency pulses will pass through a circuit. If the gate is open, then

Let's integrate the signal for $N_0$ clock pulses in time dt; $t = N_0\,\partial t$

$$V_1 = B\int_0^{N_0\,\partial t} I\,dt = B I N_0\,\partial t \qquad \text{(B.1)}$$

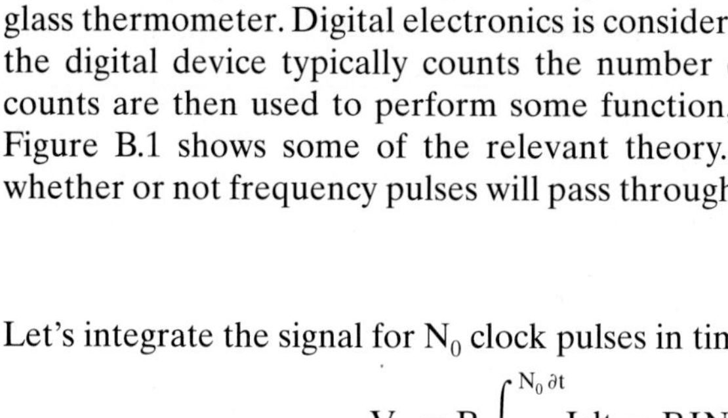

$V_2 = V_1 - B\int_0^{N_0\,\partial t} I_{Ref}\,dt = V_1 - B I N_0\,\partial t$ ($V_1$ is the signal-input; $V_2$ is the output; $I_{Ref}$ is the reference current which is obtained at $t = 0$; N is the total number of pulses.)

$$\begin{aligned} V_2 = 0 \Rightarrow V_1 = B I_{Ref} N\,\partial t \Rightarrow N &= V_1 (B I_{Ref}\,\partial t)^{-1} \\ &= B I N_0\,\partial t (B I_{Ref}\,\partial t)^{-1} \\ N &= N_0 (I/I_{Ref}) \end{aligned} \qquad \text{(B.2)}$$

**Fig. B.1.** *Schematic and relevant theory for a counter or integrator of $N_0$ clock pulses.*

the pulses pass and appear unchanged at output (compared to at input). The most common gates are termed logical gates, namely, "and," "or," "not," "nor," and "equivalence." Logic gates use transistors to sense the logic, with high logic level = 1 (usually associated with 2.4 V or greater), or low logic level = 0 (usually associated with 0.8 V or smaller).

The electronic device that accomplishes the conversion from analog to digital form is termed an analog to digital convertor (ADC). There are a few items that must be considered before using an ADC. Such as, how many bits (i.e., number of 0's or 1's) does the convertor use? Is the convertor unipolar or bipolar? Are the data of a single sign, either plus or minus (i.e., unipolar), or does the ADC use both plus and minus data (bipolar)? What is the algorithm used by the ADC to perform the data conversion? Is it a binary conversion, offset binary, 2's compliment or a BCD (sign/magnitude) conversion?

For example, a unipolar ADC that handles n bits, and has a binary code would produce a digital number, N, from an input voltage, $V_{in}$, compared to a reference voltage, $V_{ref}$, by the following formula:

$$N = [V_{in}(V_{ref})^{-1}]2^{n}, \tag{B.3}$$

with a least significant bit = $(2^{n})^{-1}$ and a corresponding code range of 0 to $(2^{n} - 1)$.

---

**Example B.1**

If a 10 bit unipolar ADC with binary code for conversion of input voltage, $V_{in}$, has a reference voltage of 10 volts, and an input voltage of 1.234 volts. Then what is the digital number that results? Round the digital number to an integer.

$$N = [1.234 \text{ Volts}(10 \text{ Volts})^{-1}]2^{10} = 0.1234\,1024 = 126.36 \approx 126$$

Now what is the binary coding for decimal 126? (01111110). Try it! (Hint: Take the first digit on the right of the binary code, namely 0, and multiply it by $2^{0}$, then add the product of the next digit to the left, namely 1, times $2^{1}$, to it, and then add the product of the next digit to the left, namely 1, times $2^{2}$, to it and so on until you run out of digits.) For the more advanced reader, what is the binary coding for 6000?

---

A bipolar ADC with a binary code has the following conversion expression,

$$N = [V_{in}(V_{ref})^{-1}]2^{n-1}. \tag{B.4}$$

Digital voltmeters or temperature-measuring devices with digital displays use what is known as a dual slope ADC. The dual slope ADC compensates for variations in both the reference and input voltages. Once the analog input from the measuring device is converted to digital value it may be displayed and/or stored. If multiple instruments are making measurements, such as with a portable meteorological tower, then the analog data from these instruments are sent to a datalogger, where they are converted to decimal, stored, and available for display, or later retrieval. The procedures for retrieving data from dataloggers are manufacturer specific, but are generally straight forward when using the manufacturers' datalogger software.

APPENDIX C

# Derivation of Wind Vane Response to Wind Direction Fluctuations

Consider a slightly underdamped vane, that is initially held at an angle to the horizontal windfield and then released. The mechanical wind vane (Fig. C.1) consists of a fin (airfoil) and a counterweight supported at opposite ends of a horizontal arm. At its center of gravity, the arm is attached to a freely-rotating vertical shaft, such that the length between the counterweight and the shaft is termed the arm length of the counterweight (r′) and that between fin and the shaft is termed the arm length of the tail (r). The air flow causes the arm to maintain a position with the counterweight into the wind, although assumed to oscillate about the true wind direction. The following is primarily derived from Wang (1975), Fritschen and Gay (1979), and Brock (1984).

The dynamics of this wind vane at a deflection angle <15° may be approached from the point of view of the forces acting on the vane. The wind exerts a force over the arm length of the tail, r, about its pivot point. This is by definition torque, †. The torque that results from the 15° deflection is (e.g., see Wang, 1975)

$$\dagger = c\,\rho\,\mathrm{Re}\ r[0.5(C_{\dagger} + C_r)b]u^2\sin\theta, \tag{C.1}$$

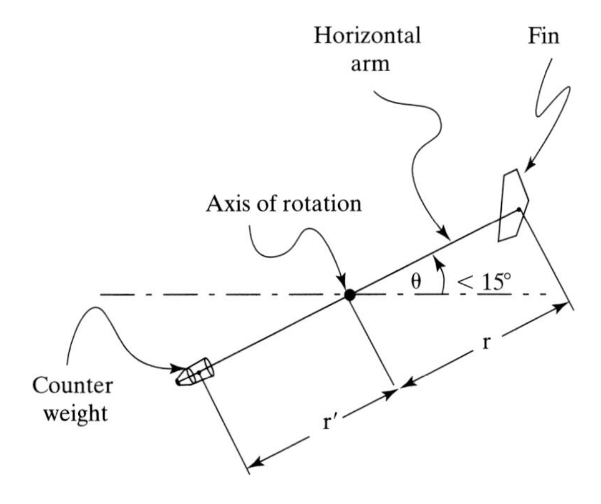


**Fig. C.1.** *Schematic of a wind vane.*

where $[0.5(C_\dagger + C_r)b]$ is the tail platform area, $A_{tp}$; b is the tail span (i.e., the fin's vertical length); $C_\dagger$ is the horizontal length of the fin that extends from the side closest to the axis to the point where a rectangle may be completed, also termed the tip chord; $C_r$ is the maximum length of the fin, also termed the root chord, $\theta$ is the instantaneous angular deflection; $\rho$ is the air density; Re is the Reynolds number; and u is the wind velocity. r and $A_{tp}$ are constants for a given vane. Assuming windspeeds between 5 and 30 mph and an angular deflection less than 15°, then $\rho$ and $R_e$ are essentially constants without any significant error. The result is a simplified torque expression

$$\dagger = K\theta,$$

where the angle of deflection has been assumed to be small, and K is a constant. The torque is, by definition, also related to the moment of inertia, $I_I$, and time, t, as

$$\dagger = I_I d^2\theta/dt^2. \tag{C.2}$$

As the force acts to decrease $\theta$, then the previous equation is more correctly written as

$$\dagger = -K\theta.$$

Then, without considering damping effects,

$$\dagger = I_I(d^2\theta/dt^2) = -K\theta$$

i.e.,

$$I_I(d^2\theta/dt^2) + (K\theta) = 0. \tag{C.3}$$

However, the damping effect or damping torque, $\dagger_d$ (i.e., the reduction of oscillation of the vane by increasing the windspeed) is directly proportional to the product of the angular speed of the rotating vane, $d\theta/dt$, and the arm of the tail, r. Hence

$$\dagger_d = ar(d\theta/dt), \tag{C.4}$$

where a is the constant of proportionality. If we let $F' = ar =$ the damping factor, then

$$\dagger_d = F'(d\theta/dt).$$

The term on the right hand side of the latter equation is the damping effect, and it should be added to our differential equation describing the dynamics of the vane motion. This yields

$$I_I(d^2\theta/dt^2) + F'(d\theta/dt) + (K\theta) = 0,$$

or (C.5)

$$d^2\theta/dt^2 + [(F'I_I^{-1})(d\theta/dt)] + [(KI_I^{-1})\theta] = 0.$$

The equation of the response of a wind vane initially held at an angle to the horizontal windfield and then released to the true horizontal position may be represented by

$$\theta = \theta_0 e^{-(t\lambda^{-1})} \sin[\phi + (2\pi t T^{-1})], \tag{C.6}$$

where T is the period of oscillation and t is the time. $\lambda$ and T depend on the dimensions and moment of inertia of the vane and the windspeed. $\lambda$ is essentially analogous to the lag coefficient of a thermometer.

The differential equation describing the dynamics of the vane motion may also be expressed in the following general form,

$$d^2\theta/dt^2 + [2\alpha(d\theta/dt)] + (\beta^2\theta) = 0, \text{ with roots, m, obtained as follows} \tag{C.7}$$

$$m = -\alpha \pm (\alpha^2 - \beta^2)^{0.5}$$

Assuming windspeeds between 5 and 30 mph and an angular deflection less than 15° with $\rho$, r and $R_e$ as constants, it becomes

$$d^2\theta/dt^2 + [(F I_I^{-1})(d\theta/dt)] + [(K I_I^{-1})\theta] = 0, \tag{C.8}$$

and its $\alpha = D(2I_I)^{-1}$ (D is the distance constant) and $\beta = (K I_I^{-1})^{0.5}$, and K is a constant $= c\rho R_e r A_{tp} u^2$, with all other symbols as previously defined. Hence the three damping conditions in terms of $\alpha$ and $\beta$ are:

1. If $\alpha < \beta$, then imaginary roots, implying that the motion is underdamped and is a simple harmonic motion.
2. If $\alpha = \beta$, then real roots and they are equal, implying that the motion is critically damped and there is no overshooting.
3. If $\alpha > \beta$, then real and unequal roots, implying that the motion is overdamped and is exponential.

*Note*: Brock (1984) provides a detailed discussion of the damping conditions for vanes in response to various first and second or high order inputs.

APPENDIX D

# Outline of the Derivation of the Psychrometric Constant

The following is primarily from Fritschen and Gay (1997). Let m represent the mass of air passing the wet-bulb (Fig. D.1) in unit time such that $m_w$ is the mass of air that that becomes saturated. Consequently, the heat loss by the m grams of air to the wet-bulb equals the heat absorbed by the wet-bulb in order to evaporate enough water to saturate $m_w$ grams of air. Once a steady state is reached,

$$P_a C_p h_t (T_a - T_w) m = m_w h_w L \varepsilon (e_w - e) \tag{D.1}$$

where $h_t$, $h_w$ are both a function of windspeed and are essentially ventilation coefficients; $\varepsilon = m_v/m_d = 0.622$; L is the latent heat of vaporization of water at $T_w$ (2,501 J/g at 0 °C).

$$(e_w - e) = [C_p P_a (L\varepsilon)^{-1} (m m_w^{-1})(T_a - T_w)(h_t h_w^{-1})] \tag{D.2}$$

and the psychrometric constant is

$$A = (C_p (L\varepsilon)^{-1})(m m_w^{-1})(h_t h_w^{-1})]. \tag{D.3}$$

If the ventilation is adequate (i.e., ventilation velocities > 3 m $s^{-1}$), then $(h_t h_w^{-1}) = 1$. Consequently,

$$(e_w - e) = [A P_a (T_a - T_w)]. \tag{D.4}$$

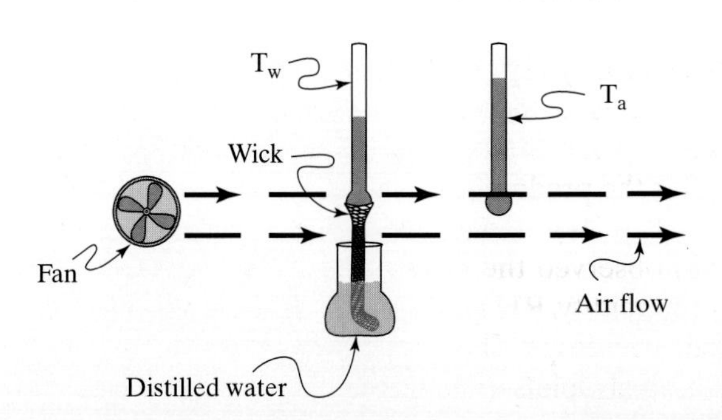


**Fig. D.1.** *Schematic of a liquid-in-glass psychrometer.*

# APPENDIX E

# Some Preliminary Field Measurement Campaign Details for an Investigation of Anatural Nuclei Concentrations During the End of a Cloud Cycle

Assume that our task is to investigate the conditions for which enhanced nuclei concentrations occur in the vicinity of evaporating cloud droplets (Plate E.1). The first thing to do is to conduct a complete literature search on our task. We could use key words like cloud evaporation, enhanced nuclei or particles, particle or nuclei generation, etc.

## *E.1 Background (From Literature Search)*

The vicinity of an evaporated or evaporating cloud has been shown to *occasionally* contain non-classically enhanced, or more than expected, nuclei concentrations compared to those present during the onset of the cloud cycle (Fig. 7.2). There have been a number of reports, since the mid-1940's, suggesting a non-classical production of nuclei in the vicinity of evaporating water surfaces, especially droplets (e.g. Dessens, 1946; Twomey and McMaster, 1955; Dinger et al., 1970; Radke and Hegg, 1972; Hegg et al., 1980; Cheng, 1988; Cheng et al., 1988; Cohen et al., 1988; Hoppel, 1988; Hegg, 1990, 1991; Hegg et al., 1990; Hoppel et al., 1990; Radke and Hobbs, 1991; DeFelice and Saxena, 1994; Saxena, 1996). The latter reports continue to grow. But, like anything else, there are those studies that have not found enhanced nuclei (e.g., Ranz and Marshall, 1952; Lodge and Baer, 1954; Blanchard and Spencer, 1964; Baumgärtner et al., 1989; Mitra et al., 1992; DeFelice and Saxena, 1994; DeFelice 1996).

Twomey and McMaster (1955) reported an apparent production of several hundred or more aerosol particles in the vicinity of evaporating sea-salt droplets during their laboratory studies when their relative humidity was below the phase transition relative humidity for the predominant salt (Twomey, 1955). Their droplets were well-ventilated, initially 1 to 10 $\mu$m in radius and crystallized around 70% relative humidity. Cheng (1988) observed the ejection of sulfate aerosols in a marine environment (25 °C, relative humidity, RH $\leq$ 75%) during the crystallization of seawater droplets derived from whitetops. This was confirmed in laboratory experiments using seawater solution droplets (Cheng, 1989). These aerosols averaged

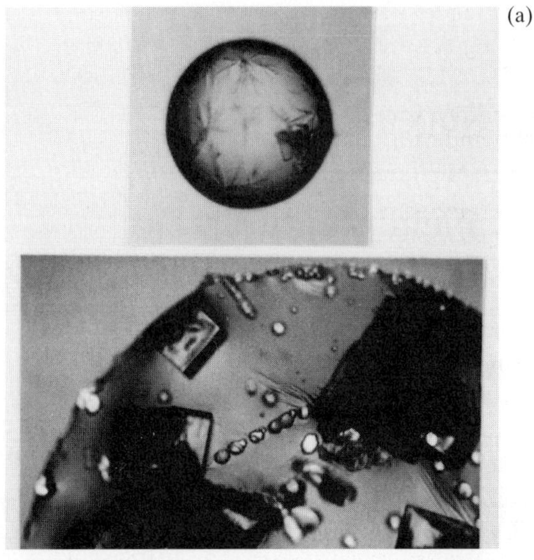


© Roger J. Cheng, ASRC, SUNYA

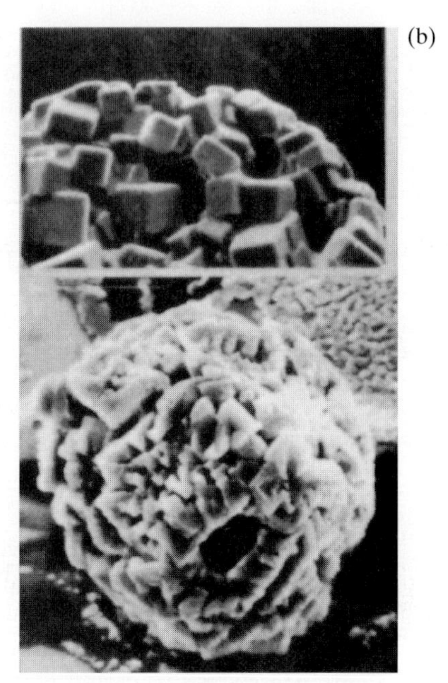


© Roger J. Cheng, ASRC, SUNYA

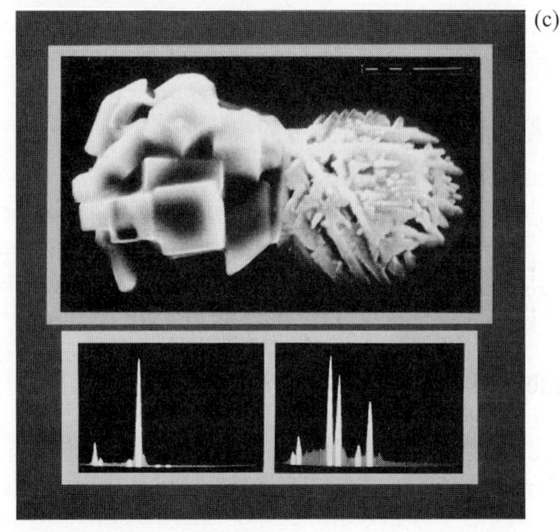


© Roger J. Cheng, ASRC, SUNYA

**Plate E.1.** *A scanning electron microscope time sequence beginning with the formation of a brine surface on an evaporating sea-salt solution droplet (a), continuing with the formation of microcrystals on the surface of this evaporating droplet (b) and ending with the no longer visible and virtually liquid water-free nuclei at the end of the evaporation process (c). The photo in (c) is only meant to illustrate the fact that the chemistry of the nuclei during the final aerosol stage (on the left) is very likely to be different than that of the escaped vapor on the right. (Photos © Roger J. Cheng, Atmospheric Sciences Research Center, State University of New York at Albany)*

$\approx$10 per droplet, and ranged from $\approx$0.1 $\mu$m to $\approx$10 $\mu$m in diameter. Ranz and Marshall (1952), and Mitra et al. (1992) found no evidence of the ejection of aerosols during evaporation, although there is no evidence that they conducted their investigation when the relative humidity was below that required for crystallization of their CCN. It is also possible that they were making their observations at a position that was too far away from the evaporating droplets to detect the ejected aerosols. Note that the detection of aerosols during evaporation was not the primary concern of Mitra et al. (1992). Nonetheless, their results are still very important, and they support the notion that relative humidity might be important. Furthermore, Hoffer and Mallen (1970), Cheng (1988, 1989) and Mitra et al. (1992) collectively indicate that the substance dissolved or incorporated into the droplet will alter its evaporation rate. Thus the composition of our evaporating droplet is likely to be of some importance. The observations of Radke and Hobbs (1991) and Hegg et al. (1990) confirm the dependence of the enhanced number of generated nuclei on humidity. Hegg et al. (1990) observed the nucleation of new sulfate particles just above the tops of marine stratocumulus clouds. Note that sulfate particles have a strong tendency of being hollow (e.g., Leong, 1981; Cheng, 1988, 1989; Mitra et al., 1992). DeFelice and Saxena (1994) suggest that environmental relative humidity and air-mass chemistry might be related to the number of cloud condensation nuclei formed during the evaporative stage of a cloud event. Saxena (1996) found four episodes of enhanced nuclei during the evaporative stage of fog at Palmer Station, Antarctica, while DeFelice (1996) did not. The microphysical and chemical characteristics of the aforementioned fogwaters are unavailable. Nonetheless, the relative humidity noted in DeFelice (1996) was above the phase transition relative humidity (Twomey, 1955) for the most probable salt(s) within the evaporating droplets. DeFelice and Cheng (1998) provide additional background details.

The literature search (include books, journals) for our task has been initiated, and it may need to be updated.

Next read the relevant literature generated from our search, and take notes. Reread Chapter 1 for guidance.

## *E.2 What Measurements Are Needed?*

For example, the preliminary evidence suggests that this non-classical generation of particles, if such is to occur or more appropriately be observed, appears to be dependent upon the: (i) environmental relative humidity and perhaps insolation; (ii) chemical composition and number density of the droplets and the interstitial nuclei throughout the period of evaporation; (iii) chemistry of the air-mass into which the evaporation occurs; and (iv) whether instrumentation can respond quickly enough and be sensitive enough to detect the changes that are likely to occur in our investigation. Insolation is added because the phenomena was observed at different latitudes and elevations (e.g., Hegg et al., 1990; DeFelice and Saxena, 1994; Saxena, 1996), and nuclei generation can occur via gas to particle conversion which may be driven by insolation. You can probably think of a couple of additional reasons yourself.

The phenomena primarily occurs during evaporation, so instrumentation needs to be capable of measuring changes in temperature, humidity, etc. with complete response times that are as fast as possible, highly accurate, and precise. The complete response times need to be measured in at least tenths of a minute if possible, and this could cost money.

The occurrence of this phenomena is not known to be prevalent, and so the campaign may initially have to include all possible events. After obtaining a better understanding of the physical and chemical processes involved, then the campaign should become more selective.

There will also be a need to consider whether making the measurements at one location will suffice or whether they be simultaneously made at another location, as well. Chances are that two sampling locations will be needed, and the separation between the two locations should be on the same length scale as the scale of our most frequent cloud trigger (i.e., are clouds that pass the site triggered by local circulations or are they triggered by synoptic?). Let us assume that there are enough resources to obtain the most sensitive, accurate instruments possible for deployment at one site.

Then we choose our measurements. What measurements are likely to be needed? Temperature, humidity, insolation, pressure, and nuclei. Why pressure? Think about it. For example, pressure is needed to determine the mixing ratio(s). Each of these sensors must respond within 0.5–1.0 s in order to have a chance at observing the generation. The nuclei concentration measurements might be able to satisfy this requirement, but present spectral measurements contain important information and require ≈12–15 s to record a spectra (DeFelice and Saxena, 1994).

The cloud is the beginning point of our task, that is, it sets the initial boundary condition for this study. The cloud droplet's final chemistry, physical, and meteorological characteristics are consequently important. The chemical component should be monitored throughout the cloud event with the greatest frequency toward its end. One of the readily available cloudwater collectors should suffice, and the chemistry of the last sample or two will be very important. Thus, one needs to know the meteorological conditions associated with the events, especially during their final stages, and how the cloudwater collectors operate.

Size-differentiated particle samples should be initiated using a multiple-stage impactor during the end of the cloud cycle. The latter helps to collect particles of different size ranges on slides or filters. The slides or filters for each size stage of the impactor may then be chemically analyzed if possible. Chances are that there will be insufficient time to collect enough aerosol mass during the actual evaporation period alone, and this collection will have to extend into the period immediately following the event, unless we can get a pump that will sample air at a faster rate than the high volume pumps used by Hi-Vol samplers (e.g. Lodge, 1994). Nonetheless, this size-differentiated chemistry information will help establish a closure to our field campaign. Gas phase analyzers would most likely give us complete closure.

## E.3 Instrument Choice

See Chapter 1, pages 3–10, and the chapters covering each of the required measurements. After the instruments are chosen and calibrated, then the data acquisition rates are determined and the datalogger programmed to query each measuring device at the optimum frequency.

## E.4 Quality Assurance

Make sure the instruments are calibrated and working properly before venturing out into the field. After setup, the calibration and operation of the instruments should be checked before beginning our measurements. Then be sure to back up the data periodically using some independent technique (e.g., write it down, copy computer-stored data to a removable diskette). Check the performance of your instrumentation twice a day, and perform a field calibration on each once a week. Instrument dependent of course, and use a primary standard whenever possible. See Chapter 1 for additional insights.

## E.5 Error Estimation

See Chapter 1, especially pages 15–20. The chapters viewed when making your instrument choices might also be helpful by providing relevant measurement accuracies, for example.

# Glossary

**absolute humidity.** This is the density of water vapor in air and is denoted as $\rho_v$. It is defined by the equation of state, $e = \rho_v R_v T$, where e is the water vapor pressure and $R_v$ is the gas constant for water vapor (461.5 J $kg^{-1}$ $°K^{-1}$).

**accuracy.** The accuracy is simply defined as how close the measured value is to a standard or unknown value, known as the "true value." The accuracy of an instrument may be found by either (i) going to the manufacturer's specifications, or (ii) calibrating it with another instrument of the same or better accuracy.

**aerosol (particle).** A solid and/or liquid particle that is primarily composed of some substance(s) other than water. This particle, however, may contain a small percentage of the thermodynamically stable bulk liquid or solid phases of water. But it is not a cloudwater drop, raindrop, water ice crystal, or the common snowflake.

**anomaly.** The deviation of a physical quantity (usually temperature, pressure or precipitation) from the normal value in a given region over a specified period. It is also the angular distance of an earth satellite, or planet from its perigee (or perihelion) as seen from the center of the earth (or sun).

**atmospheric particles.** The total set of particles that exist in our atmosphere. Actually atmospheric particles only make up about 1 part in $10^9$, or 1 ppb, by mass of the earth's atmosphere.

**audit.** An inspection of some or all components of the measurement program. Audits are either performance-based, wherein the quantitative data are independently obtained for comparison with routinely acquired data, or system-based, wherein an independent qualitative assessment is made of the program's quality control and the physical facilities and equipment for calibration and measurement (U.S. EPA, 1980).

**autocorrelation.** Primarily, but not exclusively, used in the statistical theory of turbulence which involves the study of correlations between time or spatial variant velocity fluctuations. This includes, for example, the correlation between the velocities at a fixed point at successive instants in time, or the correlation between velocities at a fixed time at different points. The correlation between velocity fluctuations which affect a particle at an instant t and at a time t + Δ later is called the autocorrelation coefficient, R(Δ). For example, if U represents the velocity at t and U′ represents the value at t + Δ later, then the autocorrelation of velocity, is gauged by the following expression for the autocorrelation coefficient

$$R(\Delta) = \sum_i (U_{i-1} U_i') \left( \sum_i U_i \right)^{-2}.$$

In this example we only have a series with one pair, and so i = 1. Normally this is not the case. An autocorrelation coefficient of 1 implies that there is no change in the time series, or that the two are the same; whereas a coefficient of 0 implies that the two are very different.

**BAROCAP®.** A small aneroid capsule with an external diameter of 35.5 mm and capacitive transducer plates inside, which is used in radiosonde units to detect atmospheric pressure. The transducer plates are supported by membranes made of a special steel alloy. Supporting rods of the plates are fixed to the membranes with hermetic glass-to-metal seals. The transducer electronics only senses the capacitance between the plates. There is no significant influence from stray capacitances between the transducer plates and the membranes, which are grounded.

**Boyle's law.** Simply, if the temperature is constant, then pressure is inversely proportional to volume. Note that if Boyle's law is combined with Charles' law (i.e., volume is directly proportional to the absolute temperature), then one can easily come up with the familiar equation of state.

**capillarity.** The effect on the reading of a mercury barometer due to a depressed meniscus (known as capillary depression) that results from surface tension. Most modern barometer scales are calibrated with the capillarity taken into account.

**cloud condensation nuclei.** The subset of condensation nuclei upon which cloud droplets definitely form.

**cloud droplet.** A bulk (meaning thermodynamically stable) liquid water droplet, formed on a cloud condensation nuclei, and growing to a size or mass that is determined by the available supply of vapor.

**condensation nuclei.** The subset of atmospheric particles upon which water vapor may condense. They are categorized by size and chemical composition, and include ice-forming nuclei and cloud condensation nuclei. The size categories are Aitken, Large, and Giant in order of increasing size. The chemical composition categories are hydrophobic, hydrophilic, and hydroscopic in order of increasing solubility in water.

**coriolis force.** An apparent force that arises from the rotation of the Earth on its axis. It primarily affects synoptic-scale and planetary-scale winds. Winds are deflected to the right of their initial direction in the Northern Hemisphere and to the left in the Southern Hemisphere.

**Dalton's law.** The total pressure in a container is the sum of the pressures of the gases in the container, and each gas exerts the same pressure it would if it were alone at the same temperature. Given a gas, such as water vapor, alone in a container of air, the pressure it exerts is called its partial pressure.

**deadband error.** The subrange through which the input varies without initiating a response, and is usually given in percent of full scale.

**deposition nucleation.** The formation of ice on a hydrophobic nuclei as water vapor comes in contact with that nuclei. If this nuclei is in a (supersaturated) vapor environment, then the ice particle will grow larger.

**dewpoint.** The temperature to which air must be cooled, with pressure (P) and mixing ratio (w) held constant, in order for the air to reach saturation with respect to water. Sometimes referred to as a dewpoint temperature, $T_d$.

**diffuse radiation.** The component of radiation that has been scattered or diffusely reflected. For example, diffuse solar radiation is defined as a non-isotropic flux of solar radiation from the sky incident on a horizontal surface and approaches a maximum as one nears the sun.

**direct radiation.** The component of radiation that is coming directly from the radiating object.

**drag coefficient.** An empirical dimensionless proportionality factor that is equal to a constant times the Reynolds number to some power. For example, if the object is spherical and placed within a uniformly moving fluid then

$$C_D = 24\,Re^{-1},$$

where Re is the Reynolds number and has a value $<1$.

**dynamic viscosity.** Simply, a proportionality factor between the time rate of change in deformation per unit length in a fluid (or just the deformation per unit length for a solid) that results when a force is applied (known as shearing stress) that is proportional to the transverse velocity gradient within a moving nonturbulent fluid. It is commonly found with units of $lbf - sec\ ft^{-2}$.

**Eulerian platform.** Generally, a measurement platform that continuously monitors certain physical parameters (e.g., wind velocity and gusts, temperature, aerosol size spectra) from a fixed platform with respect to the earth's surface.

**gates.** Gates are the basic building blocks of digital circuits. They determine whether or not frequency pulses will pass through a circuit. If the gate is open, then the pulses pass and appear unchanged at output (compared to at input). The most common gates are termed logical gates, namely, "and," "or," "not," "nor," equivalence.

**geoid.** Simply an ideal form of the earth that contains no mountains or any other topographical features, but has a similar shape.

**global radiation.** Radiation from all directions incident on a unit surface area, and includes the sum of direct radiation from all sources plus the radiation that has been scattered or diffusely reflected from/in all angles. It is usually used in conjunction with measurements of radiation. For example, global solar radiation is the total flux of solar radiation received from the solid angle of the sun's disc (known as direct solar radiation) plus radiation that has been scattered or diffusely reflected in traversing a horizontal surface of the atmosphere (known as diffuse solar radiation).

**HUMICAP®.** A thin film capacitor with a polymer dielectric sensor used in radiosondes for atmospheric humidity detection. The sensor capacitance is dependent on the water absorption in the sensor's dielectric material. The sensor is fabricated using thin film technology similar to that generally used in microelectronics.

**hysteresis.** A non-uniform needle deflection relative to an initial indication when the needle is subjected to a cyclic variation. The width of the deflection ultimately reaches a steady dimension. A related deflection, known as "creep" occurs when the quantity being measured is constant. One example of hysteresis, is a damaged or dirty needle. Hysteresis is normally associated with aneroid barometers.

**ice crystal.** A bulk (meaning thermodynamically stable) solid water particle, initially formed by the freezing of a supercooled cloud droplet or haze particle, or by deposition nucleation, and growing to a size or mass that is determined by the available supply of vapor.

**ice-forming nuclei.** The subset of atmospheric particles upon which ice crystals will form. These nuclei are typically water insoluble particles, and may also be classified as hydrophobic condensation nuclei.

**imprecision.** The remaining random error that is not repeatable and is unremovable. It can be defined by a calibration of the measuring device. The opposite of precision.

**Lagrangian platform.** Generally, a measurement platform that continuously monitors certain physical parameters (e.g., wind velocity, temperature, aerosol size spectra) while moving with the body of air that it is monitoring.

**limitation.** The capability of an instrument to give an accurate reading as long as the measurement takes place within a specific set of environmental conditions. Generally, the readings are less accurate at the extremities of the instrument's range, that is, near the endpoints of its output device. The limitation of an instrument may also be related to the span of the instrument.

**longwave hemispherical radiation.** Total hemispheric radiation minus global solar radiation.

**mixing ratio.** The mass of water vapor per unit mass of dry air, and denoted as w. Mathematically, $w = m_v m_d^{-1} = \rho_v \rho_d^{-1} = 0.622\,[e\,(P - e)^{-1}]$.

**multiplier.** A resistance connected in series with a meter movement in order to extend the range of voltages it will measure.

**net radiation.** The total downward radiation fluxes minus total upward radiation fluxes.

**operational amplifier.** The operational amplifier, OA, amplifies an input current and converts it into a voltage, where $I_{in} = I_f + I_{OA} \approx I_f$ since OA input current, $I_{OA} \approx 10^{-9} - 10^{-12}$ Amps. Furthermore, $I_{in} = (V_{in} - V_n)(R_{in}^{-1})$, $I_f = (V_n - V_{out})(R_f^{-1})$, and $V_{out} = [A\,(V_p - V_n)] + (g\,V_{cm})$, where $V_{cm}$ is the common mode voltage and $V_{cm} = (V_p + V_n)^{0.5}$; $V_p$ and $V_n$ are the positive and negative input voltages respectively; and g reflects the difference in amplification of $V_p$ and $V_n$ (i.e., the common mode gain). See Fig. G.1. The ratio of A to g defines the common mode

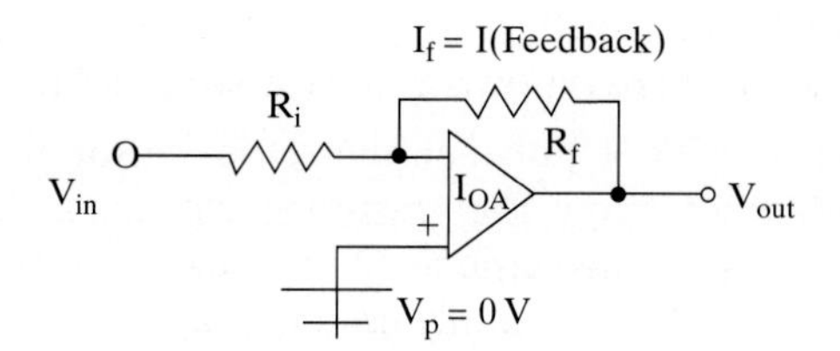


**Fig. G.1.** *An operational amplifier.*

rejection ratio (or CMRR). Generally $V_{out}$ varies between ±15 V, and typically, the amplification factor, A, is $10^5$. Some meteorological wind instruments measure current, amplify it and convert it into a voltage.

**parallax error.** Usually ≈0.1 of a scale division.

**Peltier effect.** The thermal effect due to a reversible current through dissimilar materials (regardless of whether the current was from an external source or an internal source), or through similar metals due to an external source of current. It is closely related to the Seebeck effect.

**perigee.** The point where the satellite's orbit takes the satellite closest to the object (i.e., the earth) it orbits.

**PhotoDiode.** A device that will cause current to flow when energized by light energy.

**photomultiplier.** A device wherein a resistance is connected in series with a meter whose movement is derived by light energy in order to extend the range of voltages detected by this device. It will measure low light energies.

**Prandtl number.** The measure of the relative importance of viscosity and heat conduction, or the Prandtl number, $Pr = \nu \kappa^{-1}$, where $\kappa = k\,(C_p\,\rho)^{-1}$. $\kappa$ is the thermal diffusivity or thermoelectric conductivity and $\rho$ is the density of the air.

**precision.** The variability observed among numerous measurements of a given quantity. It depends inversely on the spread or dispersion (standard deviation) of the set of measured values $X_i$ about their mean value. That is, widely dispersed, low precision (imprecision is high) and conversely. Lack of precision is caused by random error.

**primary standard.** *See* standard.

**rainfall rate.** The flux of rainfall through a horizontal surface. Sometimes called rainfall intensity. It is measured in terms of the volumetric water flux with units $m^3\,m^{-2}\,s^{-1}$, but by convention units are $mm\,h^{-1}$. Mathematically, the rainfall rate, R, can be written in terms of the diameter of the hydrometeor, D, and the hydrometeor size distribution, N(D) as

$$R = (0.16667\,\pi) \int_0^{\infty} N(D)\,D^3 u(D)\,dD$$

where u(D) is the fallspeed of hydrometeors with diameter, D. D is the melted diameter and R is the equivalent rainfall rate when the hydrometeors are snow.

**range.** The range is defined as (1) the portion of a quantity's domain that is measurable by the instrument and (2) the difference between the two extreme measurable signal readouts as indicated by the highest and lowest scale of an instrument.

**reading error.** The error obtained when reading the output; it is very common with scale instruments. Mirrors are often used to help decrease the magnitude of this error.

**reference standard.** *See* standard.

**relative humidity.** The ratio of the actual mixing ratio, w, to the saturation mixing ratio, $w_s(P,T)$, with respect to water at the same pressure and temperature. It is denoted as RH and expressed as a percentage. Mathematically, RH (%) = $100\,w\,(w_s(P,T))^{-1} \approx 100\,(e\,e_s^{-1})$.

**representativeness.** Representativeness has more to do with the phenomenon being investigated then the instrument itself. It is defined as the degree to which a measurement or set of measurements taken in a spatial-temporal domain reflects the actual environmental conditions in the same spatial-temporal domain or a different space-time domain taken on a scale appropriate to a specific application (Nappo et al., 1982).

The representativeness entails the correlation between the phenomenon being studied to the following: the spatial and temporal coverage of the sensor, and the number and distribution of the data points necessary to establish a "true" value of the environment or its component. Sensors typically sample the atmosphere on the microscale domain, although the results of these measurements are often meant to represent the meso-, synoptic-, or perhaps larger- scale atmospheric domain (Giraytys, 1970).

The number and spacing of sensors and the frequency of sampling required to give representativeness depends on the scale of the spatial-temporal domain being investigated. Consider rainfall measurements for example. How representative is a point source rainfall measurement from any gauge situated in a network of 1 gauge per 225 $mi^2$? The answer to this question requires the consideration of the type of precipitation, its intensity and duration, the effect of topography, and gauge exposure (i.e., its height, surrounding obstructions). Is the precipitation showery or continuous? The representativeness of a single point measurement for any parameter involves a temporal or spatial representativeness over very large spatial areas, or very long times (i.e., turbulence, winds, temperature, and humidity), as well as the effect of platform influences (e.g., a moving aircraft results in dynamic heating effecting temperature measurements).

**resolution.** The resolution is defined as the smallest variation in the environmental input variable that causes a detectable change in the instrument output. The resolution and accuracy usually coincide, but such is not a necessary occurrence. Note that the resolution is occasionally referred to as representativeness, or "system accuracy."

**Reynolds number.** The ratio of inertial forces to viscous forces, and is expressed as Reynolds number, $Re = U\,d\,\rho\,\mu^{-1}$, where U is the velocity of the fluid, d is the linear dimension of an object within the fluid, and $\mu$ is the dynamic viscosity of the fluid.

**secondary standard.** *See* standard.

**Seebeck effect.** The conversion of thermal energy to electrical energy. This effect "measures" the ease at which excess electrons will circulate in an electrical circuit under the influence of thermal difference. The change in the voltage (through a circuit of two dissimilar materials) is proportional to the temperature difference between the junctions when the ends are connected to form a loop.

**sensitivity.** The sensitivity is defined as (1) the change in system output for a given change of input signal; (2) the ratio of the full scale output of the instrument to the full scale input value (this is not the resolution of the instrument); and (3) the degree of accuracy an instrument output would offer.

**specific humidity.** This is defined as the mass of water vapor per unit mass of moist air, and is denoted as q. Mathematically, $q = m_v m^{-1} = \rho_v \rho^{-1} = \rho_v (\rho_v + \rho_d)^{-1}$.

**standard.** A quantity of known accuracy. There are four standards—i.e., primary, secondary, reference, and working. The highest standard is the *primary standard* followed by the *secondary standard*, the *reference standard* and finally the *working standard*.

The highest standards used by the meteorological community are referred to as secondary standards which are defined as standards that are traceable directly or indirectly to a primary standard. A primary standard is one that is calibrated with respect to the fundamental quantities, such as time, mass and length and are generally the concern of standard authorities like the National Bureau of Standards (NBS). The reference standard refers to the best standard available at a location, and should at least be a secondary standard. This should only be used if it is not practical or economically feasible to send the instrument to NBS. The working standard is calibrated against the secondary or reference standard, and is regularly used to calibrate field instruments and instruments used on a day-to-day basis. The working standard has the largest imprecision among the calibration instruments. See the National Institute of Standards and Technology (NIST) web site:

http://www.boulder.nist.gov/

for more information, including the time.

**supercooled droplets.** Liquid droplets that exist in clouds with air temperatures colder than freezing.

**surface drag force.** This term refers to the resistance or drag force acting on a body (or bodies) immersed in a uniformly moving fluid (i.e., air, water), and it is usually represented as Drag Force $= 0.5\,\rho U^2 d^2 C_D$, where $\rho$ is the fluid density, U is the velocity of the undisturbed fluid motion relative to the body, d is the linear dimension of an object with similar characteristics (e.g., shape, morphology) to the body, and $C_D$ is the drag coefficient.

**THERMOCAP®.** A sensor used in radiosondes to detect air temperature. It is based on dielectric ceramic materials. Metal electrodes are formed on both sides of a tiny ceramic chip. The capacitance between the electrodes is a function of temperature.

**threshold.** The smallest measurable input.

**Thompson effect.** The absorption or liberation of heat by a homogeneous conductor (that has a temperature gradient within it) due to a current flowing through it. It is primarily evident in currents introduced from external sources and those generated by the thermocouple itself. The ability of a given material to generate heat with respect to both a unit temperature gradient and a unit current is gauged by the Thompson coefficient.

**transducer.** A device that converts input energy of one form into output energy of another form.

**true value.** "True value" is the mean value of an *infinite* set of measurements made under constant conditions. It is *not* a *random* variable, but a constant that depends on the quantity being measured, as well as the method and apparatus of measurement. A change in the method or apparatus will probably yield a different "true" value because of a change in the systematic error. However, since it is very unpractical to make an infinite number of measurements, one usually regards the finite set of measurements made as a random sample of the infinite set. The merit of the latter should definitely be addressed in a more advanced text on this subject.

**vapor pressure.** The partial pressure of water vapor and is denoted as e.

**virtual temperature.** The temperature of dry air having the same density as a sample of moist air at the same pressure, and is denoted as $T_v$. Mathematically, $T_v \approx T[1 + (0.6\,w)]$.

**viscosity.** The result of intermolecular forces exerted as layers of a fluid (such as air and water) attempt to slide by one another. That is, it is the resistance or friction within fluids. For example, within an incompressible moving fluid, and expressed in cartesian coordinates (i.e., x, y, z) they are represented by $\partial^2 u/\partial x^2, \partial^2 u/\partial y^2, \partial^2 u/\partial z^2$ times the dynamic viscosity.

**wet-bulb temperature.** The temperature to which air is cooled by evaporating water into it at a constant pressure, and is denoted as $(T_w)$.

# Index

Boldface numbers indicate a glossary entry.

## U



## V



## W